BookWare Companion Series ™

AUTOMATIC CONTROL

The Power of Feedback
using MATLAB®

Books in the BookWare Companion Series™

BookWare Companion Series ™

AUTOMATIC CONTROL
THE POWER OF FEEDBACK
using MATLAB®

Theodore E. Djaferis
University of Massachusetts, Amherst

Brooks/Cole
Thomson Learning™

Pacific Grove • Albany • Belmont • Boston • Cincinnati • Johannesburg • London • Madrid
Melbourne • Mexico City • New York • Scottsdale • Singapore • Tokyo • Toronto

Sponsoring Editor: Bill Stenquist
Marketing Manager: Nathan Wilbur
Marketing Assistant: Christina DeVeto
Editorial Assistant: Shelley Gesicki
Production Editor: Mary Vezilich
Production Assistant: Stephanie Andersen
Cover Design: Denise Davidson
Print Buyer: Jessica Reed
Printing and Binding: Webcom Ltd.

MATLAB and PC MATLAB are trademarks of The MathWorks, Inc. The MathWorks, Inc. is the developer of MATLAB, the high-performance computational software introduced in this book. For further information on MATLAB and other MathWorks products—including SIMULINK™ and MATLAB Application Toolboxes for math and analysis, control system design, system identification, and other disciplines—contact The MathWorks, Inc. at 24 Prime Park Way, Natick, MA 01760 (phone: 508-647-7000; fax: 508-647-7001; email: info@mathworks.com. You can also sign up to receive the MathWorks quarterly newsletter and register for the user group.

Macintosh is a trademark of Apple Computer, Inc. MS-DOS is a trademark of Microsoft Corporation. BookWare Companion Series is a trademark of Brooks/Cole.

For more information, contact:
BROOKS/COLE
511 Forest Lodge Road
Pacific Grove, CA 93950 USA
www.brookscole.com

For permission to use material from this work, contact us by
Web: www.thomsonrights.com
fax: 1-800-730-2215
phone: 1-800-730-2214

Printed in Canada

10   9   8   7   6   5   4   3   2   1

Library of Congress Cataloging-in-Publication Data
Djaferis, Theodore Euclid.
   Automatic control : the power of feedback using MATLAB / Theodore E. Djaferis
      p.   cm. – (BookWare companion series)
   Includes bibliographical references
   ISBN 0-534-37171-X
   1. Feedback control systems Mathematical models. 2. MATLAB.
I. Title II. Series: PWS BookWare companion series.
TJ216.D53 1999   99-16223
629.8'3—dc21   CIP

To my parents

Euclid and *Eftychia Djaferis*

and my daughters

Jennifer and *Katherine Djaferis*

ABOUT THE SERIES

"The purpose of computing is insight, not numbers."
–R. W. Hamming, *Numerical Methods for Engineers and Scientists*,
McGraw-Hill, Inc.

It is with this spirit in mind that we present the BookWare Companion Series.™

Increasingly, the latest technologies and modern methods are crammed into courses already dense with important theory. As a result, many instructors now ask, "Are we simply teaching students the latest technology, or are we teaching them to reason?" We believe that these two alternatives need not be mutually exclusive. In fact, this series was founded on the belief that computer solutions and theory can be mutually reinforcing. Properly applied, computing can illuminate theory and help students to think, analyze, and reason in meaningful ways. It can also help them to understand the relationships and connections between new information and existing knowledge and to cultivate problem-solving skills, intuition, and critical thinking. The BookWare Companion Series was developed in response to this mission.

Specifically, the series is designed for educators who want to integrate computer-based learning tools into their courses, and for students who want to go further than their textbook alone allows. The former will find in the series the means by which to use powerful software tools to support their course activities without having to customize the applications themselves. The latter will find relevant problems and examples quickly and easily available and will have electronic access to them. Important for both educators and students is the premise on which the series is based: that students learn best when they are actively involved in their own learning. The BookWare Companion Series will engage them, provide a taste of real-life issues, demonstrate clear techniques for solving real problems, and challenge them to understand and apply these techniques on their own.

To serve your needs better, we are continually looking for ways to improve the series. Toward that end, please join us at our BookWare Companion Resource Center website:

http://www.brookscole.com/engineering/ee/bookware.html

You can recommend ways to make the series even better, share your ideas about using technology in the classroom with your colleagues, suggest a specific problem or example for the next edition, or just let us know what's on your mind. We look forward to hearing from you, and we thank you for your continuing support.

Bill Stenquist	Publisher	bill.stenquist@brookscole.com
Shelley Gesicki	Editorial Assistant	shelley.gesicki@brookscole.com
Nathan Wilbur	Marketing Manager	nathan.wilbur@brookscole.com
Christina DeVeto	Marketing Assistant	christina.deveto@brookscole.com

TABLE OF CONTENTS

PREFACE

We are literally surrounded by natural and man-made systems that are automatically controlled. A multitude of functions in the human body operate without our conscious intervention, keeping us alive and well. On a daily basis we encounter man-made systems that operate automatically and may have electrical, mechanical, chemical, hydraulic, financial or ecological characteristics. In most instances, we are not even aware of their automatic operation until there is some malfunction. A quick reflection will certainly convince the reader of the importance of automatic control in nature. It is also true that automation played a major role in the development of our highly complex technological society in the past and will continue to do so in the future.

Automatic control is a fascinating field of study! The theory and practices developed over the years provide solutions to a wide range of automation problems, giving the field its universal character. It is for this reason that courses are taught on the subject in practically every engineering department, at both the undergraduate and the graduate level. It is also true that the solution of many of these automatic control problems involves a number of disciplines, making the field truly multidisciplinary. Almost exclusively, automatic control principles and practices are offered to advanced undergraduates. However, it is important that engineering students be exposed to "engineering" early in their education in order to get motivated and excited. Automatic control is an ideal choice for an introduction to engineering. Of course, care must be taken to ensure that the material presented is at the appropriate level. One can also argue that automatic control provides the perfect setting for teaching mathematics in an engineering context. Not only can such a course provide a solid educational experience but it can be made interesting and exciting.

Automatic Control: The Power of Feedback is the textbook for an innovative introduction to automatic control. It assumes that students have had one calculus course and have basic knowledge about differentiation and integration. Automatic control has many facets and there are many pieces to the automatic control "puzzle." The design of an "automatic controller" is only one of these components. In this book we focus on the basic principle of *feedback* and show how it is used to design automatic controllers. Our approach is *model-based* because our first priority is to develop an explicit model for the system under study. For many systems these models are expressed as linear differential equations with constant coefficients. We show how to solve such simple equations *analytically* and *numerically*. Numerical computation of solutions can also be done by digital computers. We demonstrate this process using *SIMULINK*, a computer software package. We introduce the notion of system *stability* and talk about system *performance*.

We show how feedback can be used to both stabilize a system and improve performance. Theory, simulations and experiments are used to develop simple automatic controllers for a computer-controlled model car (CIMCAR-1).

We are fortunate that over the last decade or so a number of software tools have been developed that greatly facilitate the teaching and practice of automatic control. One of the most powerful, versatile and widely used packages is MATLAB/SIMULINK®, a product of The MathWorks, Inc. All the numerical computations and plotting in this book were done using MATLAB/ SIMULINK.

Here at the University of Massachusetts in Amherst this book is being used for ENGIN191, a first-year second-semester course in engineering. ENGIN191 is a college-wide course with modular structure in which each department offers a number of modules. This book is the text for a five-week Automatic Control Module. The Automatic Control Module has been offered by the departments of Electrical and Computer Engineering, Mechanical and Industrial Engineering and Chemical Engineering.

In any book-writing endeavor, the author is only one of the components that make it a success. I would like to express my appreciation to the Technology Reinvestment Project and the Engineering Academy of Southern New England for making available TRP funds for the development of this book as well as the design and construction of the prototype CIMCAR. Thanks also go to Corrado Poli, Tom Blake and Joe Goldstein for their support in securing this funding. Very special thanks go to my students Marc DiCicco and Scott Davenport for their commitment, work ethic and persistence. They put a lot of effort into designing and building the prototype CIMCAR. It was a pleasure working with them on this project. Thanks also go to Erik Ydstie for his helpful comments. I would also like to express my gratitude to The MathWorks, Inc., for software donations, and to Rick Spada and John Ciolfi, who arranged them.

Theodore E. Djaferis

Amherst, Massachusetts

1 INTRODUCTION

1.1 Automatic Control

Even though many of us do not understand the intricate details of automatic control, we are nonetheless all fascinated and excited by what can be accomplished by it. The ability of some systems to perform functions on their own, without any human intervention, has had a dramatic impact on our lives and has captured our collective imagination. Automatic control played a pivotal role in sending a man to the moon and has provided the central theme for many a science fiction novel throughout the years. It is a fact that the scientific advances of the last one hundred years have made possible the automatic operation of a multitude of systems that ensure increased productivity, promote economic development and improve the quality of life. It is also true that the demands for automation placed on us by our highly complex technological society will continue to grow during the twenty-first century and we must be prepared to meet them.

To a large extent, our lives on this earth depend on systems that operate automatically. When we say that a system operates automatically, we mean that it does so without the constant and direct intervention by a human being. Many such systems can be found in nature; others are man-made. These systems can be biological, mechanical, electrical, chemical, financial or ecological, to name just a few categories. Consider as an example the human body. It is full of systems whose continued automatic operation is essential for our existence. Think of the automatic system that is responsible for maintaining body temperature at 98.6° F, the system that regulates the heartbeat, or the focusing mechanism of the eye. Similar comments can be made for the operation of the kidneys, liver or lungs. These and many other systems of the human body operate automatically without any conscious intervention on our part. We are also literally surrounded by man-made systems that operate automatically. We come in contact with and use many of them on a daily basis. In a modern house, the interior temperature is controlled automatically by a thermostat, as is the temperature of hot water in the storage tank. The cruise-control system in a car automatically maintains a set speed, the antilock braking system automatically ensures no skidding on wet road surfaces and the emissions controller regulates exhaust emissions for cleaner air. The autopilot "flies" the aircraft for a great portion of a typical flight. An elevator dispatcher automatically sends cars to pick up passengers in a large office building or hotel. These are but a few examples of systems that operate automatically and to this list each of us can add many more.

It is natural to inquire about how this automation is accomplished. Of course, the question is quite broad and one can rightly speculate that there are many facets to the automatic control solution. Furthermore, it would be rather ludicrous to expect that a complete answer to the problem can be provided in an introductory text like this one. However, a great deal of insight about automatic control can be obtained by limiting discussion to some basic principles and focusing the investigation on specific problems. Therefore, our objective in this introductory book is to help the reader gain a better understanding of the fundamental issues involved and formulate solutions to some specific automatic control problems.

Control deals with the automatic operation of dynamic systems. One can think of a "dynamic system" as an object that is "excited" by external "inputs" and produces "outputs." The word *dynamic* implies change and variability and consequently involves the notion of "time." The word *system* is used to describe a very broad range of objects with different characteristics. At a conceptual level, systems can be represented by "blocks" that have "inputs" and "outputs." This is the so-called *input-output* description (see Figure 1.1).

Figure 1.1
A Conceptual Block Diagram of a System

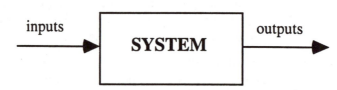

Consider, for example, an "electric oven." We can think of this as a system with one input, the voltage applied to the heating element, and one output, the temperature of the chamber. Another example is a water storage tank. We can think of the tank as the system, the water supply as the input and the water level in the tank as the output. The brake system in a car can also be thought of as a dynamic system, with the car as the system, the position of the break pedal as the input and car speed as the output. A conceptual block diagram of this system is shown in Figure 1.2.

Figure 1.2
Conceptual Block Diagram of a Car Brake System

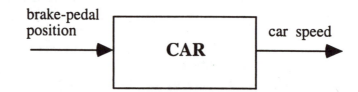

Let us explore the notion of "dynamic system behavior" a little further. Consider the brake system presented above and suppose a vehicle is moving at a certain speed down a deserted highway. We can press the brake pedal say by .5 cm and hold it there for a certain period of time (this is the input) and observe what happens to the vehicle speed (which is the output). We all know that the speed will steadily decrease until the vehicle comes to a complete stop. The way the vehicle responds will depend on its dynamic characteristics. Given the same input (i.e., brake pedal is pressed and held in exactly the same way) a car, a pick-up truck and a trailer-truck will respond quite differently.

The dynamic characteristics will also be different, if the vehicle is fully loaded or empty and if the road surface is wet, dry or icy.

Manual operation of a system implies direct and constant intervention by a person. To complete tasks, human beings rely on their basic sensing, actuating and reasoning capabilities. In order to have the same system operate automatically, all these functions must now be performed by physical devices like *sensors*, *controllers* and *actuators*. Sensors are used to measure quantities, controllers generate the appropriate control commands and actuators implement them. Our scope in this book is limited and therefore we will focus attention on only one component of the automation problem, namely the controller. We assume that appropriate actuators and sensors have been designed and constructed and are available for our use.

It is true that even though systems have diverse characteristics (e.g., electrical, mechanical, chemical), in many cases their operation can be described by mathematical models that are surprisingly similar. This realization allows one to proceed and investigate the operation of these systems in a unified way and to provide solutions that can be applied universally. This approach to automatic control is called *explicit-model* based because the first priority is to develop explicit mathematical models that describe the operation of dynamic systems. Invariably, these models are only approximate representations of system behavior. Nevertheless, having these mathematical descriptions or models of system operation facilitates the process of both system analysis and controller design. Specifically, consider the task of automating the brake system for a car. In order to automate this function and replace the driver in this activity, one would need to design the required controller (as well as introduce appropriate sensors and actuators). Having a mathematical description or model of how the speed is affected by pressing the brake pedal allows one to efficiently develop the required controller.

The development of mathematical models that describe the dynamic operation of systems is an important first step in our approach to automation. It is not difficult to explain what this means. In the example of the car brake system, the input is the position of the brake pedal over some time period (say $0 \leq t \leq T$). Well mathematically, we can think of this as a function of time. To be specific, assume that the position of the brake pedal is the function $u(t) = .5$ for $0 \leq t \leq T$ and zero otherwise. The corresponding car speed can also be thought of as a function of time, say $y(t)$. A model for the dynamic behavior of the car brake system is just a mathematical expression that relates the input to the output. For any particular input $u(t)$ it gives us in a direct or indirect way the corresponding $y(t)$. Now, we could have applied a different input and the response of the system would also be different. For example, rather than pressing the pedal abruptly by .5 cm, we could have pressed it gradually in some fashion. Clearly, the manner in which the car speed would respond would then be different. Usually this mathematical model does not provide an exact expression of the actual output, but only an approximation. In other words, the car speed computed from the prescribed mathematical model does not exactly match the observed car speed from an actual experiment. However, this does not prevent one from using this model to analyze system behavior and design appropriate controllers for automatic operation.

For many systems this approximate mathematical description (model) of system behavior is a single differential equation that relates the input and the output. Typically, the input shows up as the function on the right side of a differential equation and the output is its solution. This means that one can study the dynamic operation of a system by studying the solution of differential equations. One can also use this mathematical model to suggest algorithms (schemes, strategies) for automatic control (i.e., the development of controllers). As a consequence of these models, controllers are frequently expressed as adjustable parameters in an "overall system" differential equation.

In this book we exploit this model-based approach, introduce basic principles and present methods for automatic control (i.e., show how to construct controllers). Since there will be a system model available, one can employ three different "tools" in order to solve the problems encountered: theory, simulations and experiments. Let us explain in more detail what this means. Suppose that our task is to develop an automatic controller for a dynamic system and that the operation of the overall system is described by a differential equation. Assume that the controller is an "adjustable" parameter in this differential equation.

- For a specific controller, one can use basic theory to solve the corresponding differential equation and obtain analytical expressions (formulas) for the solution. Furthermore, one can examine how different controllers effect the solution.

- One can solve these same differential equations "numerically" by computer (i.e., by running a computer program) and display the results. This process is called *simulation*. If the numerical computation is accurate, then the outcome of a digital computer simulation should agree with the exact calculation (i.e., analytical solution).

- One can implement the controller on the actual system, perform experiments and collect measurements of the output. These results can be compared with the corresponding data obtained from the analytical computation and the digital computer simulation. If the model and digital simulation are accurate, then analytical solutions, simulations and experimental results should all be in agreement.

This model-based approach shows how theory, simulation (digital computer computation) and experiments are used together in order to gain knowledge about system behavior and generate solutions to automatic control problems.

There are many examples of how the introduction of automation increases productivity, facilitates system operation and improves the quality of life. Rather than discussing issues at an abstract level, it would be very helpful if we considered a specific example. Since automatic brake systems for cars are not yet widely available, we will concentrate on a very familiar vehicle system, namely, *cruise-control*. Let us suppose that we were given the task of developing a cruise-control system for a car (i.e., "automating" the speed control function of the driver in highway driving). There are several other operations that a typical driver has to perform manually in order to successfully drive a car on the highway. These include keeping the car in the lane, avoiding collisions with other vehicles and changing lanes, but let us just concentrate on speed control and how this process can be automated. Automatic speed control systems–cruise-control–are now taken for granted, but there was a time not too long ago when this was "science fiction."

We are all able to appreciate and are intrigued by the automatic operation of some system or machine, but few of us perhaps have taken the time to inquire in more detail as to how this is done. This is certainly true for the cruise-control system in a car. We can see that when we set the device to 55 mph on the highway, car speed is controlled automatically and we don't have to worry about pressing the gas pedal (as long as we do not touch the brake). How is this achieved automatically? Indeed, there are many important implementation issues that must be addressed and several conceptual issues that have to be worked out, but it turns out that the basic principle behind the solution of this problem is not difficult to explain.

Before we present the solution framework for this problem, it would be very instructive to first see how a human would execute the same task *manually*. After all, the automatic system must replace the functions of the human driver. To simplify matters, let us assume that we are on a deserted highway, that the road is flat and that we will control the speed by using just the gas pedal (throttle). We will not make use of the brake, as is currently the case with existing cruise-control implementations. The typical driver looks at the speedometer periodically and checks the speed. If the speedometer reads 45 mph, then the driver pushes down the gas pedal and the car accelerates. The driver eases up when the car speed reaches 55 mph. Similarly, if the speedometer reads 60 mph, the driver reduces pressure on the gas pedal until the car slows down to 55 mph. The car slows down because of tire/engine friction and air resistance. It is evident that the driver performs four actions (functions):

- determines the actual car speed by looking at the speedometer

- generates an "error" by comparing the actual car speed with the desired speed of 55 mph

- decides what is the proper action to take depending on this error

- implements this action by adjusting the throttle via the gas pedal

It is instructive to visualize all this in a conceptual block diagram, which shows specific functions and identifies pertinent variables (see Figure 1.3).

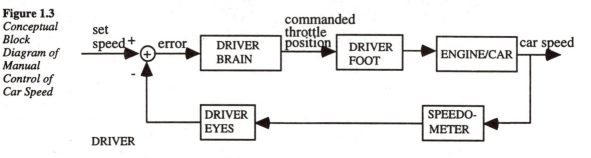

Figure 1.3
Conceptual Block Diagram of Manual Control of Car Speed

Some of the lines in this diagram are identified with relevant variables (i.e., quantities like car speed or commanded throttle position). It is helpful to think of the lines connecting the blocks as carrying information and arrows as indicating the direction of information flow between blocks. The blocks represent system components that perform functions and can be viewed as receiving/processing inputs and generating outputs. Specifically, the engine/car block takes the throttle position (input) and turns it into car speed (output). In other words, if the throttle is at a certain position (which implies a certain gas pedal position) the car will travel at a certain speed. There are many other car components including the cooling system, the brakes, the battery and the electrical system, and variables or quantities such as oil pressure, engine temperature, brake fluid pressure and alternator voltage present in the system that we do not need to worry about. We are only concerned with the functional relationship between throttle position and car speed. Therefore, we can ignore those other system components and focus solely on the problem at hand. The actual car speed is measured by the car speedometer, which displays it to the driver. The driver's brain then takes this information as well as the set speed, which is kept in mind, processes it and decides what to do with the gas pedal so that the error is made equal to zero (i.e., car speed is made equal to the set speed). The typical driver does not think about all these things, as she/he has learned to do them almost subconsciously. However, it is important to be able to identify these functions if we are to replace the human driver by devices that perform them automatically.

What is immediately apparent in the above description is that the driver uses the fundamental principle of *feedback* in order to perform the task of regulating the car speed. Information on the actual car speed from the speedometer is "fed back" and compared with the set car speed, which resides in the driver's brain. This error is used to determine what action to take (command) in order to make the error smaller. If the error is zero, which implies that actual car speed is the same as the desired car speed, no corrective action is required.

If we want to make this system operate automatically, without human involvement, it makes sense to employ the same idea. We would like to replace all the functions of the driver with other devices that perform them automatically. Let us first address the function of the "speedometer" block and the "driver eyes" block and replace their function with an appropriate device (i.e., a sensor). Since "eyes" will not be used to convey this information to the brain, a different device must be used that measures the speed and expresses it as a quantity that can be used (say a voltage signal). The speedometer should be replaced by a different *speed sensor*. Assume that we have such a device available, one that represents 0 mph as 0 volts and 100 mph as 10 volts, with a linear scale for the values in between. This means that 55 mph would correspond to 5.5 volts. In this eventuality, the set or desired speed would also need to be expressed in *volts* for consistency and so the error will be in *volts*.

The next block that must be replaced is the one labeled "driver brain." It receives the error signal and based on that decides what the correct control action should be. This function will now be performed by a man-made device that is referred to as the *controller*. It is instructive (and factual on many occasions) to think of this device as being a microprocessor (mini-computer). Humans have learned how to perform this task after months and years of driving. The microprocessor does not have this "driving experience" and must be programmed correctly to take the appropriate action.

Let us now focus more carefully on what should be the appropriate action to be programmed on the microprocessor (i.e., develop the controller). Let us suggest the following simple control scheme, inspired by how a human would act: every five seconds, check the error and do one of two things. If the error is positive (set speed more than actual), press throttle down a specific amount (say .5 cm) and hold it there. If error is negative (set speed less than actual), pull throttle back a specific amount (again .5 cm) and hold it there. This control scheme (algorithm) can be very easily programmed on the microprocessor. It is a very simple algorithm that seems to do the right thing. Of course, one can question how well it will work. Even though there is a clear distinction between the microprocessor, which is a device, and the control algorithm, which should be thought of as software, we frequently refer to both as the controller.

Finally, we also need to introduce another device, the *throttle adjuster*, which will carry out the required throttle movement (i.e., execute the command provided by the controller). This device is the *actuator* and we assume that an appropriate one has been provided. A conceptual block diagram showing the information flow and the specific devices involved in an automatic control implementation of cruise-control is shown in Figure 1.4. As you can see, the diagram looks very similar to Figure 1.3, which shows at a conceptual level how manual speed control is achieved.

Figure 1.4
Conceptual Block Diagram of Car Cruise-Control System

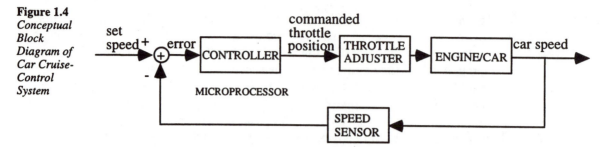

We have just suggested a very simple scheme for controlling car speed. The controller solution presented is rather simplistic and is not the one used in current cruise-control implementations. However, this feedback framework is precisely the one used.

In our earlier discussion we mentioned how important it is to develop mathematical models for the system that accurately describe its operation, and yet our control solution did not require such a model for the car. However, we questioned how well the proposed control scheme would work. Our intention in this example was not only to showcase the feedback solution method, but also to stress the importance of using a model. Let us elaborate on this issue in the context of answering the question of how well the proposed control scheme would work.

Suppose that the car actually used is a pick-up truck and assume that the control scheme suggested performs well when the truck is empty. Now, let us load up the truck. More power (gas) will be needed to accelerate the truck as desired and it would take longer to reduce truck speed. So the control scheme would need to be readjusted. Suppose next that on the day we first tested the controller there was no wind blowing, but that on the next day we had a strong head wind. We again recognize that if all worked as desired the first day, the system would not operate as well with a strong head wind and that the control scheme would again need to be readjusted. On the contrary, if we were able to develop a mathematical model for our system, car weight and the wind disturbances would have been explicitly included. A *robust* controller could then have been designed that worked well in a variety of situations.

We used the cruise-control example to present some basic ideas behind automatic control. It should be evident that this approach and method of solution can be applied to a vast number of systems and a multitude of applications. It has helped us identify some of the important issues and basic elements involved in a solution of an automatic control problem. Even though it was relatively easy to explain the whole process, there is a great deal of work to be done to design and implement an automatic speed regulator system that performs well. Several questions are immediate: How does one design and build an appropriate speed sensor? Does one need to have the car speed reported continuously, or is it acceptable to obtain a speed measurement periodically, say every second? How are two voltage signals subtracted? What device can do this? We reported one control algorithm, but how good is it? Will it meet the performance specifications that we want? What should the performance specification be anyway? Is it acceptable for the error to be ± 1 mph, or do we really want it to be zero? How fast should the correction be done? Should we allow "jerky" motion and not worry about passenger comfort? What are some safety concerns? What happens if we have a malfunction in any of the components? How should we design the actuator (throttle adjuster)? How much will the system cost? Clearly, since we have cruise-control systems in cars, these issues have been addressed satisfactorily for this application. Even though we will not be able to answer many of these questions in this book, we will nonetheless identify important topics and lay down a foundation for future study.

1.3   A Basic Introduction to Automatic Control

Because of the overwhelming presence of automatic control all around us, it is not at all surprising to see interest in the subject from students, practitioners and researchers from a broad range of fields. This is particularly true in engineering, where regardless of which specific branch one chooses (aerospace, chemical, civil, electrical, industrial or mechanical) one can find control applications. Typically, there are courses offered in each department that present basic principles, develop theory and discuss how automatic control can be implemented. In fact, one can say that automatic control is the only engineering discipline that is found in every engineering department. This gives automatic control a rather unique and "universal" character.

Traditionally, undergraduate students are first exposed to automatic control toward the end of their program, as time must be spent in developing the appropriate mathematics, physics and chemistry skills. These introductory control courses are then followed by several others at the graduate level. However, this need not be the case, as fundamental principles and basic control concepts can be used to motivate students much earlier. The material can be presented in an interesting and exciting way, allowing the student to gain a clearer understanding of what engineering is all about. A typical second-semester freshman has enough knowledge and sophistication to understand the basic concepts, appreciate the engineering problems and formulate simple solutions. As the necessary background is obtained from mathematics, physics, chemistry and other engineering courses, one can then revisit the topics, delve more deeply into the subject matter and gain a more complete understanding. Having a first-year "motivational" course in control is only one of the reasons it should be offered early in the program. Another reason is that automatic control provides the perfect setting for teaching mathematics in the context of engineering.

It should be clear from our earlier discussion that the problem of automatic control is very basic, interdisciplinary and multi-faceted. It is unrealistic to expect to cover all the different aspects and learn all the required material at this introductory level. Our objective in this book is to present some basic engineering principles and fundamental concepts about automatic control. We will also formulate and solve some basic automatic control problems.

As mentioned, we will focus on only one of the pieces of the automatic control "puzzle," namely the design and operation of the controller shown in Figure 1.4. Recall that we used the word *controller* to refer to both the control algorithm and the device that implements it (i.e., the microprocessor that runs the control algorithm). More specifically, in this course we will focus on the control algorithm itself and show how to develop control algorithms so the system meets performance specifications. We will assume that all the other pieces of the puzzle, which include actuators and sensors, have been addressed and operate correctly.

In the cruise-control example we suggested a very simple control algorithm. Namely, if the error was positive, regardless of how large, the control action was to push the throttle by a certain amount, and if the error was negative, regardless of how large it was, to pull back the throttle a certain amount. Let us think a little more about this control scheme. Checking the error every five seconds seems to be kind of arbitrary and so is pressing or pulling back by exactly .5 cm. It would make better sense that if the error was positive and large we should press *more* and if it was positive and small press *less*. Furthermore, the simple control strategy would make the motion of the car quite jerky and all passengers sick. We could have suggested many other control algorithms. A variation of the one just presented would be the following: Rather than pressing or pulling back by a certain amount (regardless of the size of the error) we can do it in a way that is *proportional* to the error. In other words, if the error is +10 mph (set speed 10 mph more than actual) we multiply it by a constant (e.g. .05 cm/mph) which would result in the throttle being pressed by +.5 cm. If the error is -5 mph, then the correction would be -.25 cm, which would mean that we would pull our foot back by .25 cm (which is the same as pressing down by -.25 cm). One can immediately recognize that such a scheme would have better performance characteristics. However, even this one has its disadvantages. If we choose the proportionality constant that is too small, it would take a long time for the error to become zero; if it is too large the error would oscillate about zero. In view of our earlier discussion about the need for system models, it should be evident that the proper choice of the controller gain should be based on the system model.

There are two general methodologies for developing control schemes (algorithms, strategies). The first is inspired by how humans develop control capabilities – via *learning*. Through repeated trials (which include failures) we *learn* how to perform a given task. Think of how you first learned to drive a car. You observed other adults driving cars for a number of years, you took a driver's-ed course and then you trained on an actual car. No one asked you to develop mathematical models for the car or acquainted you with Newton's Laws of Motion. Yet you were able to learn how to drive. We could have proceeded in this fashion for the development of controllers in the cruise-control example without first explicitly generating models by carrying out experiments with different controllers and observing the results. Specifically, we could have postulated that a "constant gain controller" was one to use and carried out experiments with different values for the gain. After a number of trials with different gains, we would choose the control gain that gave us the best results. The second methodology uses physical laws, like Newton's laws from physics and in some cases specific experiments (see Chapter 7) to develop explicit dynamic models (mathematical descriptions) of the systems to be controlled. These dynamic models are then used by the engineer to develop the appropriate controller. In this book we will deal exclusively with this second, *explicit-model* based approach to automatic control.

Our experience over the years shows that spending time developing such dynamic models greatly facilitates the process of designing appropriate controllers. Dynamic models are just mathematical descriptions of relationships between system variables. Specifically, for the cruise-control example the system to be controlled is represented in Figure 1.3 by the block labeled "engine/car." The dynamic model is a mathematical description of the relationship between throttle position and car speed. In particular, suppose the car is standing still with the engine running and we put it in gear and push down the throttle a certain amount. What will happen to the car speed? We know from experience that the car will accelerate and reach a cruising speed. We would like to develop mathematical expressions (equations, formulas) that give us this relationship. This is not as hard as it may initially seem to be. A very convenient such representation for this system and many others can be given using *differential equations*. This process of developing dynamic models for systems is referred to as *modeling,* which, as mentioned, is an important first step in efficient controller design.

With regard to modeling, we need to make two important observations. Even though we try to generate mathematical descriptions that accurately describe the operation of the physical system, these will always be approximations. After all, this is why we call these descriptions *models*, because they seldom are exact representations. It turns out that working with approximations rather than exact descriptions is quite appropriate for control. The second relates to the use of the word *dynamic*, which as pointed out earlier implies variability or change (opposite of *static*). In the cruise-control example, when the throttle is pressed down and held by a certain amount, car speed changes (increases) until it reaches a certain level. This takes time and does not happen instantaneously.

Furthermore, we are not just interested in its final value, but the manner in which speed reached its final value as well. It is important to us, the engineers, to know if there were sharp changes, oscillations or other such variations. We are interested in how speed changes (i.e., in the "dynamics"). So too are the passengers in the car, who will "feel" all these effects and will not be very happy if the ride is "bumpy."

Our first priority in this book, will be a discussion on system modeling (Chapter 2). We already mentioned that these dynamic system models are frequently given in terms of differential equations. We will first discuss briefly how physical laws lead very naturally to the development of dynamic models in terms of differential equations. We will show this by considering simple mechanical, electrical and chemical systems. Our next priority will be the discussion of very simple differential equations and their solution (Chapter 3). In the cruise-control example, the ability to develop dynamic models in terms of differential equations and then solve them does allow us to obtain a description of car speed as a function of time. Remember, this would be the speed of the model and not the car itself, but if our model is reasonably accurate, this would be pretty close to the actual car speed. We will give analytical solutions to differential equations (i.e., develop mathematical expressions) but also show how one can get a digital computer to calculate these solutions using some numerical techniques. In other words, we will enter the necessary data into a computer and the computer will calculate the solution. We will not have to write our own software to do this, but will use a very popular software package called **MATLAB/SIMULINK** (Chapter 4). As mentioned, we use the word *simulation* to describe the process of computing the solution of a differential equation numerically on a digital computer.

We have already pointed out that the operation of a dynamic system can be described, in many cases, in terms of differential equations. One can solve (hopefully) these differential equations and find exactly how the system operates *quantitatively*. From a different perspective, we are also interested in characterizing the operation of a dynamic system in a *qualitative* manner. In other words, we do not want to know exactly how the system operates, but only basic characteristics of the operation. One such characteristic is that of *stability* (Chapter 5). We all have an intuitive notion of what is a stable system. It is one that in normal operation does not have variables whose values "blow up." Temperatures remain bounded, pressures remain bounded, velocities remain bounded, and so forth. There are many different definitions of stability, all revolving around this intuitive notion of "boundedness." Stability is perhaps the primary performance objective, but others exist as well. Suppose that a system is stable (i.e., system variables are well behaved, no blow ups). Then we may want to know the characteristics of the output of the system, when it is excited by an input that is a step function. In the context of our cruise-control example, this is precisely the speed response of the car when we push down and hold the throttle by a fixed amount. Important characteristics include: (1) the speed of response of the system, (2) the time it takes for the final speed to be reached, (3) the existence of overshoot in the speed response and (4) the presence of oscillations.

We already know from the manual control of systems that feedback is frequently used to ensure the successful completion of a task. Upon closer examination one will discover the presence of feedback in a wide range of situations. In the context of automatic control, feedback can be used to dramatically change the dynamic characteristics of a system. In particular, it can be used to make an unstable system stable and to improve system performance. In this text we will see exactly how feedback is used to improve step response performance (Chapter 6).

In a typical control application, the engineer is given the task of developing an automatic controller for a specific physical system. A standard solution to such a problem includes dynamic modeling, controller design and analysis (based on this model) computer simulation and, finally, hardware implementation and experimental testing on the actual system. As we conclude our presentation, we would like to demonstrate this process by considering a specific example. The application we have chosen involves a computer-controlled model car, CIMCAR-1, in a collision avoidance task (Chapter 7). We will develop a differential equation model, suggest controllers, compute analytical solutions, perform simulations and finally conduct experiments with CIMCAR-1. In doing so, we demonstrate how the approach proposed in this book, which exploits theory, simulations and experiments, does solve automatic control problems.

2 SYSTEM MODELS AND DIFFERENTIAL EQUATIONS

2.1 Models of Simple Mechanical Systems

Physicists, chemists, biologists and many other scientists have been studying natural phenomena for centuries. In many instances they were able to discover through observations and experiments that a number of physical processes obey basic physical laws. These physical laws are expressed in terms of mathematical expressions that involve physical parameters, which provide complete descriptions of system behavior. Some of the best known are Newton's laws of motion, which deal with the motion of objects in certain environments. The area of physics that studies these issues is called classical mechanics. Kirchoff's laws and Ohm's law govern the behavior of electrical circuits. The laws of conservation of mass and energy determine the operation of chemical systems. In this chapter we will confine our discussion to simple mechanical systems, electrical circuits and chemical systems. We will see how basic laws can be used to derive mathematical descriptions (*models*) of the dynamic behavior of such systems.

Consider an object that has mass m and is constrained to move in a straight line, say the x-axis. Suppose that a force f is acting on the object along the x-axis. Newton, through experimentation, was able to determine that such an object will begin to move along the x-axis and in fact accelerate with acceleration a. He was able to show that force, mass and acceleration are related by the now famous formula:

$$f = ma \qquad (2.1)$$

(i.e., force is the product of mass and acceleration). For a specific object, the mass is usually a constant but the force and acceleration are quantities that may vary with time. So it is more appropriate to show this time dependence by writing $f(t)$ and $a(t)$. The acceleration $a(t)$ is the rate of change of velocity $v(t)$ with respect to time, and velocity is the rate of change of position $x(t)$ with respect to time. If $x(t)$ denotes the position of this object as a function of time, then $x(0)$ is the position of the object at time $t = 0$, and $x(3)$ is the position of the object at time $t = 3$, $v(0)$ is the velocity at time $t = 0$ and $a(1)$ is the acceleration at time $t = 1$. Since the velocity is the rate of change of position with time, we can express this relationship using derivatives. We can write:

$$\frac{dx(t)}{dt} = v(t) \qquad (2.2)$$

Furthermore, since the acceleration is the rate of change of velocity with time we can write:

$$\frac{dv(t)}{dt} = a(t) \qquad (2.3)$$

Differentiating both sides of expression (2.2) with respect to time and substituting in (2.3), we obtain a relationship between the second derivative of position with respect to time and acceleration, namely:

$$\frac{d^2x(t)}{dt^2} = a(t) \qquad (2.4)$$

Therefore, Newton's (second) law of motion can also be written in perhaps a not so familiar form as:

$$m\frac{d^2x(t)}{dt^2} = f(t) \qquad (2.5)$$

This expression tells us that if a force $f(t)$ is acting on this object of mass m in the direction of motion (along the x-axis), then its position $x(t)$, as a function of time, must obey this relationship. This is a *differential equation* because it involves derivatives. For the system composed of the mass m moving in a straight line and being acted on by the force $f(t)$, equation (2.5) gives an exact mathematical description of its dynamic behavior. We can certainly think of this as an "input-output" description for this system, with the force being the input and the position of the object being the output. An input-output block diagram for this system is shown in Figure 2.1.

Figure 2.1
Input-Output
Block Diagram

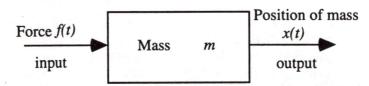

The system behavior is dynamic, not static, because for some specific force the position and perhaps the velocity and acceleration of the mass change with respect to time. Our discussion of Newton's law was rather abstract because we did not specify a particular system or identify the origin of force $f(t)$. We can apply this law to develop mathematical descriptions or models for a great variety of mechanical systems. Let us begin with the system depicted in Figure 2.2, which shows a block of mass m on a horizontal table with a frictionless surface, which is constrained to move in a straight line, the x-axis.

Figure 2.2
Mass on
Frictionless
Surface

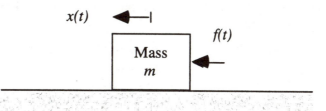

We push the block with an external force $f(t)$ and $x(t)$ (positive direction is that of the arrows) is the position of the center of mass of the block, measured from some reference point. The direction of motion is the x-axis and along this axis the only force acting on the block is $f(t)$. We have assumed that the surface is horizontal, so that the force of gravity is not in the direction of motion, but rather perpendicular to the direction of motion. So applying Newton's law to this particular system will provide us with a mathematical description (i.e., a dynamic model) of the movement of the block:

$$m\frac{d^2x(t)}{dt^2} = f(t) \tag{2.6}$$

Our assumption that the table is positioned in the horizontal plane, which implies that the gravity force will not affect the motion of the block in the x direction, is quite reasonable. However, our assumption about the surface being frictionless is only appropriate for a limited number of cases. If we try to push a book on the kitchen table with some force $f(t)$, we will feel a friction force that resists motion. At this point we will not pursue the development of a differential equation for a system with this type of friction, but defer discussion of this topic for later. Let us consider next the system shown in Figure 2.3.

Figure 2.3
Mass-Spring
System

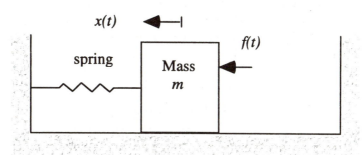

We have a block of mass m that is connected to a spring and is allowed to move on a horizontal surface. The spring, in turn, is connected to the wall. Let $x(t)$ be the deviation of the center of mass of the block from the "rest position" (i.e., when the block is not moving and the spring is not compressed). The positive direction is toward the wall (direction of the arrows). We make the following assumptions for this system:

A1) The block is constrained to move along the *x*-axis on a frictionless surface.

A2) If the spring is compressed (or stretched) by an amount *x(t)*, then it will exert a force $f_s(t) = -kx(t)$ on the block (this is known as Hooke's law) where *k* is a positive constant. If $x(t)$ is positive (spring compressed) this force will be in the negative direction (opposite to arrows) if $x(t)$ is negative (spring stretched) this force will be positive. Mass of the spring is negligible.

Now suppose that we also apply an additional external force *f(t)* on the block. We would like to combine all this and develop a dynamic model for the system. In other words, we would like to develop a differential equation (we expect this in view of Newton's law) that describes its operation. We consider the external force *f(t)* to be the input to the system and the position *x(t)* to be the output. We are interested to know how this block will move if we, say, pull it or push it with our hand (which means that *f(t)* has some nonzero value for a period of time) and then let it go (which means that the value of *f(t)* is eventually set to zero). Now, one might argue about whether the assumptions made about the spring reflect reality, but let us for the moment assume these to be true. We have a mass, the block, that is constrained to move along the *x*-axis and there are two forces acting on it, *f(t)* and $f_s(t)$. From Newton's law we have that the total force *F(t)* acting on the block in the direction of motion is $F(t) = f(t) + f_s(t)$, which implies that the mathematical description, or dynamic model for this system becomes:

$$m\frac{d^2x(t)}{dt^2} = f(t) - kx(t) \qquad (2.7)$$

This implies that if we are given *f(t)* and are able to solve this differential equation for *x(t)*, we know exactly how the block moves with time. In other words, given the input *f(t)* we can compute the output *x(t)*. Our next example is the system depicted in Figure 2.4.

Figure 2.4
Spring-Dashpot-Mass System

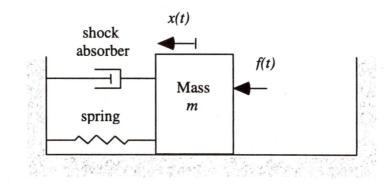

One more component was added to the mass-spring system, a "shock absorber" also called *dashpot* or *piston*. Let $x(t)$ again be the deviation of the center of mass of the block from the "rest position" (i.e., when the block is not moving and the spring and shock absorber are uncompressed). Let us also make one additional assumption:

A3) The shock absorber exerts a force $f_p(t) = -pdx(t)/dt$ on the block, where p is a positive constant and its mass is negligible.

Proceeding in exactly the same manner as above we can write:

$$m\frac{d^2x(t)}{dt^2} = f(t) - kx(t) - p\frac{dx(t)}{dt}$$

which can be rewritten as:

$$m\frac{d^2x(t)}{dt^2} + p\frac{dx(t)}{dt} + kx(t) = f(t) \tag{2.8}$$

Again, we can think of this mathematical model as an input-output functional description of our system. For this spring-dashpot-mass system, we can think of the force $f(t)$ as being the input to the system and the position $x(t)$ of the block as being the output. So the dynamic system takes a specific input, operates on it and generates a corresponding output. A block diagram showing this input-output relationship is shown in Figure 2.5.

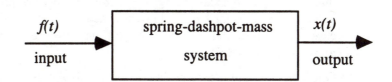

In the previous examples we used Newton's law of motion as it applies to the motion of objects that are constrained to move in a straight line. There is also a version of this law that applies to objects that rotate about some axis. Specifically, it states that the moment of inertia J of an object, multiplied by the angular acceleration $d^2\theta(t)/dt^2$, is equal to the algebraic sum of the torques $\tau(t)$ that act on the object about the axis of rotation. Specifically:

$$J\frac{d^2\theta(t)}{dt^2} = \tau(t) \tag{2.9}$$

The moment of inertia J is a parameter that depends on the mass and mass distribution of the object about the axis of rotation. Let us apply this law and develop the model of a the single link robotic manipulator shown in Figure 2.6.

Figure 2.6
*Single Link
Robotic
Manipulator*

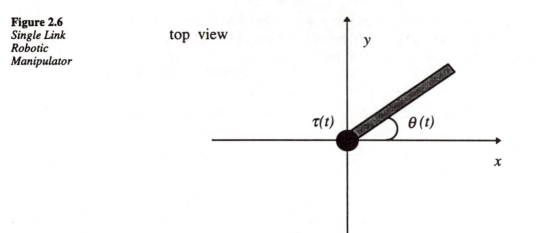

The single link robotic manipulator operates in the horizontal plane and can rotate about the z-axis, which is perpendicular to the page. The torque $\tau(t)$ is externally applied to the link (positive direction is counterclockwise) and the link rotates with angle $\theta(t)$, measured from the x-axis in a counterclockwise manner. If we assume that there is no friction at the joint (i.e., the only torque applied to the link is the external torque) then expression (2.9) is a mathematical model that describes the operation of this system.

In these examples we looked at specific mechanical systems, which are made up of components that are assumed to behave in the manner prescribed. In particular, we assumed that the "spring" obeyed Hooke's law, the force due to the dashpot has a specific form and friction effects were negligible. It may be that a specific physical spring obeys this law only approximately, or that the value of the constant k is not known exactly, or that friction is present. In that case the differential equation model we just developed will only be an *approximate* dynamic description.

One can use the physical laws presented above to develop mathematical models of quite complicated mechanical systems like six degree-of-freedom robotic manipulators. Such are the machines we frequently see on television commercials, performing automatic spot-welding or painting operations on car assembly lines or doing pick-and-place operations in other automated manufacturing facilities. The same methodology is also used in the development of mathematical models for rockets, planes and ships.

2.2   Models of Simple Electrical Systems

Newton's laws describe how objects move when they are subjected to forces or torques and can be applied to mechanical systems. These are perhaps the most well known to the reader from the study of high school physics. There are many other physical laws that govern the behavior of other types of systems like electrical, chemical, hydraulic or aerodynamic. Kirchoff's current and voltage laws apply to electrical circuits. Electrical circuits are interconnections of electrical components like resistors (think of them as lightbulbs) inductors (solenoids, coils) and capacitors (two parallel metal plates separated by air). If components like these are connected to each other and then to some voltage source (such as a battery), then they form a closed circuit and electric current will flow. This is what happens when we load up a flashlight with batteries, we push the switch and the light turns on. We can use a circuit diagram to show this: the lightbulb is represented by a resistance R, a positive constant, which is connected to a battery. The battery supplies a voltage $v(t)$, a current $i(t)$ flows in the circuit (see Figure 2.7) and the light comes on.

Figure 2.7
Simple Circuit

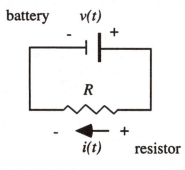

What actually happens is that the battery sets up an "electric field" in the wire and electrons that are negatively charged particles "feel" a force that makes them move. This is much like the force exerted on masses that are in a gravitational field. Electrons move in the circuit from low potential (negative battery port), to high potential (positive battery port). However, we would like to think of this movement of electric charges, referred to as the *current*, in terms of the motion of positively charged particles. Positive charges move the opposite way, from high potential to low potential. This is the direction of the current shown in Figure 2.7.

Just like in the case of mechanical systems, there are physical laws that govern the dynamic behavior of electrical devices. The flow of current in the resistor obeys Ohm's law, which states that:

$$v(t) = Ri(t)$$

That is, the voltage across a resistor is proportional to the current through it, the proportionality constant being the value of the resistance. The voltage-current relationships for other circuit components can also be developed. The one for an inductor, a coil of inductance L (a positive constant) is given by:

$$v_L(t) = L\frac{di_L(t)}{dt}$$

where the voltage across an inductor is proportional to the rate of change of current through it. The one for a capacitor of capacitance C (a positive constant) is:

$$i_C(t) = C\frac{dv_C(t)}{dt}$$

That is, the current "through" a capacitor is proportional to the rate of change of voltage across it. These individual components can be connected with each other to form electrical circuits. If a battery, an inductor (denoted by L and the appropriate circuit symbol), a resistor (denoted by R and the appropriate circuit symbol) and a capacitor (denoted by C and the appropriate circuit symbol) are all connected in "series," we obtain the circuit shown in Figure 2.8.

Figure 2.8
An RLC Circuit

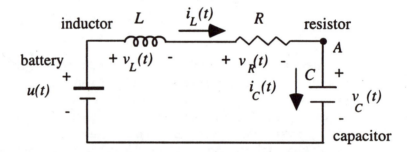

Using Kirchoff's laws we can write down mathematical expressions that describe the operation of this circuit. Given the voltage $u(t)$ of the battery, these expressions allow us to compute the current that flows in the loop and the voltages across each of the other components.

Kirchoff's current law states: *the sum of the currents that flow into a node (a connection of circuit components) is equal to the sum of the currents that flow out of the node.* Specifically, in Figure 2.8 the current that flows into node A (denoted as $i_L(t)$) is equal to the current that flows out of the node (denoted as $i_C(t)$). Therefore:

$$i_L(t) = i_C(t) = C\frac{dv_C(t)}{dt}$$

Kirchoff's voltage law states: *the algebraic sum of the voltages around a closed loop is equal to zero.* Applying this around the loop in Figure 2.6, we have:

$$v_L(t) + v_R(t) + v_C(t) - u(t) = 0$$

Using these two expressions and the voltage-current relationships of individual elements allows us to write:

$$L\frac{di_L(t)}{dt} + Ri_L(t) + v_C(t) = u(t)$$

which implies:

$$\frac{d^2 v_C(t)}{dt^2} + \frac{R}{L}\frac{dv_C(t)}{dt} + \frac{1}{LC} v_C(t) = \frac{1}{LC} u(t) \qquad (2.10)$$

This electrical circuit is a reasonable model for an interconnection of physical components (a resistor, inductor, capacitor and voltage source). This differential equation will therefore be a good mathematical model, albeit approximate, of the operation of such a physical system. If we place voltmeters and ammeters, the values we would measure would be close to those predicted by this mathematical model. More about circuits can be found in the book by Bobrow (1987), listed in the bibliography.

2.3   Models of Simple Chemical Systems

Just as Newton's laws of motion allowed us to derive mathematical models for mechanical systems and Kirchoff's laws did the same for electrical circuits, the fundamental principles (or laws) of the conservation of energy and mass allow us to develop dynamic models for chemical systems. Specifically, we know from physics that "energy may be transformed from one kind to another, but it cannot be created or destroyed." In other words, energy is conserved. Similarly, mass can be neither created nor destroyed in some process, which implies that mass is conserved. In nuclear physics we know that mass and energy are very much related ($E = mc^2$) and therefore cannot be considered separately. However, for the types of systems we will be considering, it is appropriate to think of conservation of energy and mass independently. These conservation principles can be expressed in a variety of ways and can be used to develop mathematical models for chemical systems.

Let us apply these laws to investigate the operation of an isothermal continuous-stirred-tank reactor shown in Figure 2.9. Suppose that we know from chemistry that the product B can be produced from some material A by the reaction $A \rightarrow B$ with reaction rate r. This reactor is used to produce the product B from raw material A.

Figure 2.9
*An Isothermal
Continuous-
Stirred-Tank
Reactor*

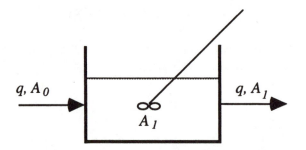

The reactor is basically a tank filled with fluid, which has an inflow and an outflow. A fluid feed mixture, which contains the reactant A in some concentration, is supplied to the tank from the inflow at the rate q. In the tank the reaction $A \rightarrow B$ takes place and material B is produced, having some concentration. Reactant A and product B leave the tank from the outflow. The outflow rate of fluid is also q. We are interested in the manner in which material B is produced. Specifically, we want to develop an expression that tells us how the concentration of material A in the tank (and outflow) changes with time. Knowing the concentration of A allows us to directly compute the concentration of product B. Let us assume that A_1 denotes the concentration of material A in the tank and outflow stream. Let the concentration of material A in the inflow be denoted by A_0. We make the following assumptions about the operation of this chemical system: (1) As indicated, the flow rates of liquid into the tank and out of the tank are the same, equal to q, and are constant during the operation. (2) The volume V of liquid in the tank is constant. (3) The temperature of the liquid during operation remains constant (isothermal process). (4) The tank is continually stirred and concentrations throughout the tank and in the outflow stream are the same. (5) The reaction rate, $A \rightarrow B$, is proportional to A_1, i.e., $r = kA_1$, where k is a constant.

For this chemical system, we can make the following observations: (1) material A is being supplied to the tank from the inflow and (2) some amount of material A is used to produce material B, another portion accumulates in the tank and some flows out. We can certainly apply the conservation of mass principle as it relates to material A and state:

$$accumulation = input - output$$

In particular, let us consider what happens in a suitably small time interval Δt, from time t to time $t + \Delta t$. Since the concentration of material A in the tank may change with time, let $A_1(t)$ be the concentration of material A at time t. Furthermore, in the general case the concentration of material A in the inflow can also change with time, so we denote it as $A_0(t)$. We have assumed that the volume of liquid in the tank is constant and so the amount of material A in the tank at time t will be $VA_1(t)$. Then $A_1(t + \Delta t)$ is the concentration of A at time $t + \Delta t$ and $VA_1(t + \Delta t)$ the amount of A at time $t + \Delta t$. So the accumulation of material A in the tank during the time period Δt is $VA_1(t + \Delta t) - VA_1(t)$. The amount of material A coming into the tank in time Δt is $qA_0(t)\Delta t$. The amount of material A that "leaves" is $qA_1(t)\Delta t$ (in the outflow) and $kVA_1(t)\Delta t$ (because of the reaction). Since mass is conserved we must have:

$$VA_1(t + \Delta t) - VA_1(t) = qA_0(t)\Delta t - qA_1(t)\Delta t - kVA_1(t)\Delta t$$

If we divide both sides by Δt and take the limit as $\Delta t \to 0$, we can write:

$$V \frac{dA_1(t)}{dt} = \lim \Delta t \to 0 \ \frac{A_1(t+\Delta t) - A_1(t)}{\Delta t} = qA_0(t) - qA_1(t) - kVA_1(t) \quad (2.11)$$

Rearranging terms, this results in:

$$V \frac{dA_1(t)}{dt} + (q + kV)A_1(t) = qA_0(t) \quad\quad (2.12)$$

We immediately see that the concentration of material A in the tank satisfies this differential equation (see Douglas, 1972).

Suppose now that we have two identical continuous stirred tank reactors in series as shown in Figure 2.10, where the outflow of tank 1 is the inflow of tank 2. In both reactors we have the reaction $A \to B$ taking place. The concentration of material A in reactor 1 is A_1 and in reactor 2 it is A_2.

Figure 2.10
Two Continuous Stirred Tank Reactors in Series

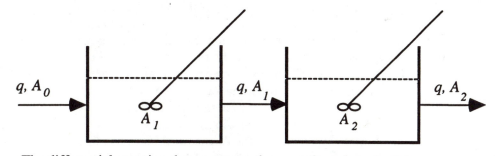

The differential equation that expresses the operation of reactor 1 is:

$$V \frac{dA_1(t)}{dt} + (q + kV)A_1(t) = qA_0(t) \quad\quad (2.13)$$

Repeating the same arguments as above for reactor 2 leads to the differential equation:

$$V \frac{dA_2(t)}{dt} + (q + kV)A_2(t) = qA_1(t) \qquad (2.14)$$

We can manipulate these two equations in the following manner. First, multiply equation (2.13) by q. Since q is a constant we have:

$$V \frac{dqA_1(t)}{dt} + (q + kV)qA_1(t) = q^2A_0(t)$$

Next, substitute expression $qA_1(t)$ from equation (2.14) into this equation.

$$V \frac{d(V\frac{dA_2(t)}{dt} + (q + kV)A_2(t))}{dt} + (q + kV)(V \frac{dA_2(t)}{dt} + (q + kV)A_2(t)) = q^2A_0(t)$$

Performing the appropriate simplifications we end up with:

$$V^2 \frac{d^2A_2(t)}{dt^2} + 2V(q + kV) \frac{dA_2(t)}{dt} + (q + kV)^2A_2(t) = q^2A_0(t) \qquad (2.15)$$

Note that the concentration of material A in reactor 2 satisfies a differential equation. More about chemical system modeling and control can be found in Douglas (1972) and Ogunnaike and Ray (1994).

It is obvious from the above discussion that the operation of mechanical, electrical and chemical systems, which are governed by their respective physical laws, can be modeled by mathematical expressions (differential equations), which look exactly the same. Specifically, compare equations (2.8), (2.10) and (2.15). If someone gave us just these equations with specific values for the different parameters, we would not be able to distinguish which equation came from which system. This continues to be true for other types of systems as well. We find that in many cases, system models for different types of physical systems (electrical, chemical, aerodynamic, mechanical, etc.) can be developed that have the form of linear differential equations with constant coefficients (as equations (2.8), (2.10) and (2.15)). This is a very interesting observation for at least two reasons:

- if we learn how to solve such differential equations, then we in essence know how to determine and analyze the dynamic behavior of a wide range of systems

- since the input-output descriptions of many systems are the same, if we learn how to develop a controller for one of them, we will know how to develop controllers for all. This is a very efficient way to proceed and a great advantage of the model-based approach suggested in this book.

2.4   The Need for Solving Differential Equations

In the next chapter we will see how to compute the solution of such differential equations. Here we just would like to motivate that work. If we are to develop automatic controllers for systems, we had better know how these systems behave. If we apply some force to an object, we would like to know how it will move. If we apply some voltage to some circuit, we would like to know what are the voltages across specific devices and the currents through them. Consequently, if these systems have a dynamic behavior that is accurately described by differential equations, it is imperative that we know how to solve these equations to determine their dynamic characteristics.

Let us return to examine the implications of a system with a dynamic model given by equation (2.5). Suppose we have a block of mass $m = 1$ kg (we use the MKS system of units) and are pushing it with some force $f(t)$ on a frictionless surface. Assume that at time $t = 0$ the object is at position $x_0 = 2$ m and it has velocity $v_0 = 1$ m/sec, and that at time $t = 0$ we stop pushing it (i.e., $f(t) = 0$ for all $t \geq 0$). How will the block move? What will be its position at some specific point in time $t > 0$?

Expression (2.5) says that the second derivative of the position with respect to time (which is also equal to the derivative of velocity with respect to time), will be:

$$\frac{d^2x(t)}{dt^2} = \frac{dv(t)}{dt} = 0 \qquad \text{for all } t = 0 \qquad (2.16)$$

Since the derivative of the velocity with respect to time is zero, this implies that the velocity does not change with time (i.e., it is a constant). Since at time $t = 0$ we know that the velocity is 1 m/sec, this means that the velocity stays at this value for all $t \geq 0$. Velocity is the derivative of position with respect to time, and so we must have:

$$\frac{dx(t)}{dt} = 1 \qquad (2.17)$$

If the object is moving with a constant velocity, then its position $x(t)$ as a function of time must be expressed by:

$$x(t) = t + 2 \qquad \text{for all } t \geq 0 \qquad (2.18)$$

We can check all this out, because if we differentiate both sides of expression (2.18) once with respect to time, we have $dx(t)/dt = v_0$ and if we differentiate one more time we have: $d^2x(t) = dt^2 = 0$. Furthermore, we see that if we plug in the value $t = 0$, we have $x(0) = x_0 = 2$ m. We have just solved this differential equation! Expression (2.18) gives us a formula for the position of the mass as a function of time. In particular, at time $t = 5$ sec, $x(5) = 7$ m, at time $t = 10$ sec $x(10) = 12$ m, and so on. For this very simple system, we know exactly how the block will move.

We have just seen that if an object constrained to move in a straight line is acted on by a force, then its position as a function must obey equation (2.5). Usually we will be given the force and we will want to compute the position $x(t)$ as a function of time. Since equation (2.5) involves derivatives, it is not an algebraic equation but rather a *differential equation*. Even though the reader's experience with such equations may be very limited, there should be no reason for alarm. Differential equations are very important in mathematics and engineering. Since we do not have the necessary background, we will only be able to solve some elementary equations. However, this knowledge will be enough for us to comprehend some very basic concepts, which will help us in our work on automatic control.

The reader should already be able to see a connection between Newton's law and our discussion of automatic control in Chapter 1. Recall the cruise-control example. A car is an object, which has some mass. In order to get the car moving, there must be a force acting on it. Clearly, in this case the car engine will provide this force. From the last time we tried pushing a car to get it moving, we know that there is quite a bit of friction that we have to overcome from the road surface (let us neglect wind resistance). So there is some other force, namely friction, that opposes the force provided by the engine. The engine force is bigger because the car moves. The car's movement is governed by Newton's law. In other words, Newton's law can be used to develop a dynamic model for our system in the cruise-control application. Admittedly, the situation there is a little more complicated because we need to develop a description (dynamic model) that gives us the relationship between throttle position and car speed. We realize that the position of the throttle is directly related to the force generated by the engine, which pushes the car, but we are certainly not able, with our current knowledge, to provide mathematical expressions that describe this relationship. But if we had that relationship, we could then combine it with Newton's law to obtain an appropriate dynamic model for use in the cruise-control system.

At this point it is important to mention that dynamic models of systems can also be obtained by performing carefully designed experiments. There are ample reasons for this practice. One is the fact that in many cases it is not easy to develop mathematical descriptions by going back to first principles (like Newton's laws). For example, in the above discussion on the development of dynamic models for mechanical systems, we neglected to take into account the effects of friction, even though we acknowledged its existence. Friction effects are almost always present in mechanical systems, but it is not easy to characterize their impact directly in terms of mathematical expressions. However, the effects of friction can be taken into account by performing experiments and developing system models from the data collected. We will elaborate more on these issues in Chapter 7, when we discuss the dynamic model of a computer-controlled model car, which will be used in experiments. Another reason is that even if we could develop a dynamic model from first principles, it may be expressed in terms of very complicated mathematical relations. Developing the dynamic equations of a six-degree-of-freedom robotic manipulator is a good example. The differential equations obtained are very complicated. Simpler models can also be developed through experiments. Even though not as accurate, these simpler models are easier to work with for system analysis and the development of automatic controllers.

2.5   Exercises

Problem 1   Using Newton's law, write down the differential equation that describes the operation of the system shown below. Make assumptions A1, A2 and A3 for this system (see Section 2.1).

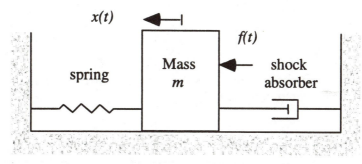

Problem 2   Using Newton's law, write down the differential equation that describes the operation of the system shown below. Make assumptions A1 and A2 for this system (see Section 2.1). Both springs are exactly the same (spring constant k).

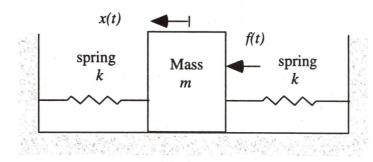

Problem 3   Using Newton's law, write down the differential equation that describes the operation of the system shown below. Make assumptions A1 and A2 for this system (see Section 2.1). The two spring constants are k_1 and k_2.

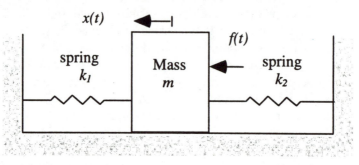

Problem 4   Use Kirchoff's laws to write down the differential equation that describes the operation of the circuit given below:

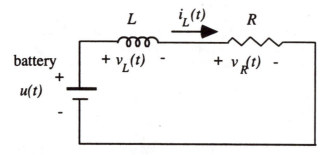

Problem 5   Use Kirchoff's laws to write down the differential equation that describes the operation of the circuit below:

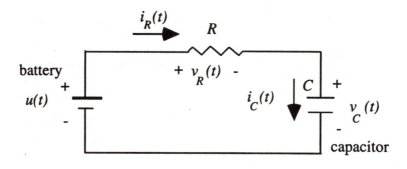

3 LINEAR DIFFERENTIAL EQUATIONS AND THEIR SOLUTION

3.1 Solving Differential Equations

We saw in Chapter 2 how differential equations arise very naturally in science and engineering. We saw this in the application of Newton's laws to mechanical objects, Kirchoff's laws to electrical circuits and the conservation of mass law to chemical systems. Application of these laws leads to the development of differential equation models for a great variety of systems. This fact is a primary reason why differential equations were studied by mathematicians and why a great deal of theory has been developed on the subject. At this introductory level, all we want to do is gain a very basic understanding of differential equations and their solution. These differential equation models will be subsequently used for system analysis and controller design.

Our discussion will focus on what are called *first* and *second order linear differential equations with constant coefficients*. These are equations that take the form:

$$\frac{dx(t)}{dt} + bx(t) = u(t) \tag{3.1}$$

$$\frac{d^2x(t)}{dt^2} + a\frac{dx(t)}{dt} + bx(t) = u(t) \tag{3.2}$$

respectively, where a and b are given constants and $u(t)$ is a given function of time. These are precisely the type of differential equations we have seen in Chapter 2, models of simple electrical, mechanical and chemical systems. These are equations that involve derivatives of functions and whose solution is not a specific value, but rather a specific function of time. Our discussion will first address particular equations and then continue with the general case.

Before we begin our discussion about differential equations, let us first consider some simple algebraic equations and compute their solution. This will also give us the opportunity to contrast their solution with that of differential equations. Consider the algebraic equation in some variable x :

$$3x + 4 = 1 \tag{3.3}$$

The obvious solution is $x = -1$. In other words, if we take the value -1 and substitute it in equation (3.3), we have that the right side is equal to the left side. Equation (3.3) does not involve "time" and so the solution is just a specific constant. Things would not be very different, however, if in such an algebraic equation in x, time was introduced represented by the parameter t. Specifically, we may want to solve for x the equation:

$$2x + 4sin(t) = 2 \qquad (3.4)$$

Since this equation does involve "time," we should also be interested in the interval of time over which we want to compute a solution. Assume the interval of interest $0 \le t \le \infty$. The solution of this equation is not a constant, but rather a function of time. We can easily see that the "x" that solves it is: $x(t) = 1 - 2sin(t)$. To see that this is the solution, we just substitute $1 - 2sin(t)$ for x in (3.4) and see that the left side is equal to the right side, for all values of t in the interval of interest, here $0 \le t \le \infty$. Clearly, the solution is no longer a constant, but is parameterized by t. Said another way, the solution is now a function of t.

First Order Differential Equations

Differential equations are different from algebraic equations because they involve derivatives. Let us look at the following simple differential equation:

$$\frac{dx(t)}{dt} = 1 \qquad (3.5)$$

This is a *first order* differential equation because only the first derivative of a function $x(t)$ appears. The first thing we have to notice is that even though $x(t)$ could turn out to be a constant, since we are differentiating with respect to t (think of t as time) we expect that in general $x(t)$ would be a function of time. We are also interested in some interval of time, say $0 \le t \le 10$, over which we would be computing the solution. To solve this equation we need to find a function $x(t)$ of time, so that when we substitute it in the equation, the left side would equal the right side, for all values of t in the interval of interest. Here we will try and "guess" the solution. We ask the question: What function of t when differentiated with respect to t is a constant equal to 1? If $x(t) = t$, then the derivative with respect to t is equal to 1, so this is certainly a candidate for the solution. However, one can easily see that there are many other functions of t that would qualify as well. Namely, $x(t) = t + 1$ is also equal to 1 when differentiated with respect to t. In fact, any function of the form $x(t) = t + a$, where a is an arbitrary constant will solve equation (3.5), since when such an $x(t)$ is substituted in (3.5), the left side becomes equal to the right side, for all t in the interval of interest. So we immediately see that we do not have a single solution to equation (3.5), but many solutions.

Let us change the problem statement a tiny bit. Suppose we want to solve the same differential equation:

$$\frac{dx(t)}{dt} = 1 \qquad\qquad (3.6)$$

over the interval $0 \leq t \leq 10$ *and* also require that $x(0) = -2$ (think of this as an "initial condition"). We refer to this as an *initial value problem*. We already know that an $x(t) = t + a$ will solve the differential equation, and since we now also require that $x(0) = -2$, this specifies that $a = -2$. Therefore,

$$x(t) = t - 2 \qquad\qquad (3.7)$$

will solve the differential equation and satisfy the initial condition (i.e., solve the initial value problem). It seems that we have identified a unique solution to this differential equation. Indeed, if the solution is of the form $x(t) = t + a$, then yes, we have found the unique solution, but it may be that there are other forms that we have not considered. It turns out that one can prove that there are no others! The proof of this fact is beyond the scope of our discussion and will be left for future study. See Boyce and DiPrima (1977) for more details.

We could have proceeded in developing a solution to this simple differential equation in a different manner, using the concept of *integration*. Equation (3.5) says that a function of time on the left,

$$x'(t) = \frac{dx(t)}{dt}$$

(where $x'(t)$ stands for derivative of $x(t)$ with respect to time) is equal to a function of time on the right (i.e., 1). Since these functions are equal, if we integrate them over some interval, say from 0 to v, where v is a value in the interval $0 \leq v \leq 10$, then the integrals must be equal. Recall that a definite integral is just the area under a curve. In other words:

$$\int_0^v x'(t)dt = \int_0^v 1dt$$

From the fundamental theorem of calculus, the left side evaluates to $x(v) - x(0)$. The right side evaluates to v. Therefore, we have that $x(v) = v + x(0) = v - 2$. Now, v is a "dummy" variable that we can change to t and denote the solution as $x(t) = t - 2$.

Let us now consider the differential equation:

$$\frac{dx(t)}{dt} + x(t) = 0 \qquad\qquad (3.8)$$

---

over the interval $0 \leq t \leq 10$. From our previous work, it would seem that in order to have a unique solution we also have to specify an initial condition (say $x(0) = 1$). If we try the form of solution used in equation (3.6), we will see that the scheme will not work. For example, let $x(t) = t + b$. Then substituting into the left side of (3.8) will give: $1 + t + b$. Regardless of what we chose for b, this will never be equal to zero (the right side) for all time in the interval. If we let $x(t) = b$, a constant, then we will need to have $b = 0$. But if $x(t) = 0$ for all time in the interval, then the initial condition would not be satisfied (unless it also happened to be zero). We need to guess a different function form. Let us rewrite equation (3.8) as:

$$\frac{dx(t)}{dt} = -x(t)$$

So now we are looking for a function of time that, when differentiated, is equal to the negative of the function. A very special function with this property is the *exponential*, $x(t) = fe^{-t}$, since $x'(t) = -fe^{-t}$. If we also want to have $x(0) = 1$, this means that $f = 1$. The solution of (3.8) with initial condition $x(0) = 1$ is given in equation (3.9) and plotted in Figure 3.1.

$$x(t) = e^{-t} \qquad\qquad (3.9)$$

Figure 3.1
A Plot of the Solution to Equation (3.8)

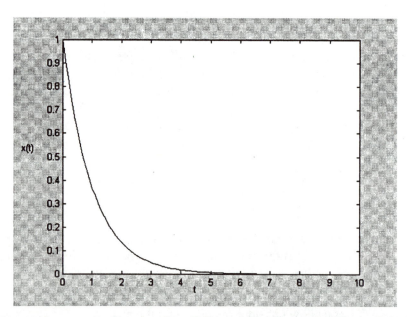

It is interesting to note that equation (3.8) could be the model for the RC circuit of Problem 5 in Chapter 2. Since $u(t) = 0$, the battery should be replaced by a "switch," which closes at time $t = 0$. Then $x(t)$ would be the voltage across the capacitor and $R = 1$, $C = 1$. Since $x(0) = 1$, the initial capacitor voltage is equal to 1. When the switch closes, current will flow and the capacitor will be "discharged" (i.e., its voltage will asymptotically go to zero).

One can see that similar reasoning leads to the solution of the "general" first order differential equation:

$$\frac{dx(t)}{dt} + qx(t) = 0 \qquad (3.10)$$

in the interval $0 = t = 10$, with some initial condition $x(0)$, where q is some arbitrary *nonzero* constant. It is very easy to verify that $x(t) = fe^{-qt}$ is a solution since $-qfe^{-qt} + qfe^{-qt} = 0$. The constant f is again determined from the initial condition by evaluating fe^{-qt} at $t = 0$ and setting it equal to $x(0)$ (i.e., $f = x(0)$). The factor "$-q$" that appears in the exponent of the exponential function can also be given a different interpretation. It is the root of the polynomial $s + q$. This polynomial in s is formed from the differential equation by "replacing" $x(t)$ by 1 and $dx(t)/dt$ by s. The polynomial $s + q$ is called the *characteristic polynomial* associated with the differential equation.

Let us now consider the situation when the right side of the differential equation is a function other than zero. Specifically, consider the differential equation:

$$\frac{dx(t)}{dt} + qx(t) = u(t) \qquad (3.11)$$

where q is a *nonzero* constant and the function $u(t)$, the *input*, is the so-called *unit step* function:

$$u(t) = \begin{cases} 0 & t < 0 \\ 1 & t \geq 0 \end{cases} \qquad (3.12)$$

The interval of interest is $0 \leq t \leq \infty$ and, in addition, we have the initial condition $x(0) = 0$. It is interesting to note that equation (3.11) could again be the model of the circuit in Problem 5 of Chapter 2. When $R = 1$, $C = 1$, the battery voltage $u(t)$ is a step function and the initial charge on the capacitor is zero, that system model is precisely equation (3.11) with $q = 1$. Since the right side of (3.11) is no longer the zero function but the step function, one can show that the solution will now be composed of two parts:

- the first part is that computed for the case when the right side is equal to the zero function (as in equation (3.10)) and is equal to $x_h(t) = fe^{-qt}$. This is referred to as the *homogeneous solution*.

- the second part, for a step function input, will be a constant $x_p(t) = a$ and is called the *particular solution*.

So the *complete solution* will be of the form:

$$x(t) = x_h(t) + x_p(t) = fe^{-qt} + a$$

The value of the constant a can be obtained by substituting $x_p(t) = a$ in the equation. For this case we have for any nonnegative t that $qa = 1$. So it must be that $a = 1/q$, which implies that $x_p(t) = 1/q$ and the complete solution will be $x(t) = fe^{-qt} + 1/q$. The constant f can now be computed by using the initial condition. In particular, $x(0) = 0 = f + 1/q$, which implies that $f = -1/q$. The complete solution of (3.11) then becomes:

$$x(t) = -\frac{1}{q} e^{-qt} + \frac{1}{q} \tag{3.13}$$

One can again show that the solution just computed is unique. When $q = 1$, a plot of the solution–the step response, as it is called–is given in Figure 3.2 for the interval $0 \le t \le 10$.

Figure 3.2
A Plot of the Solution to Equation (3.11) when q = 1

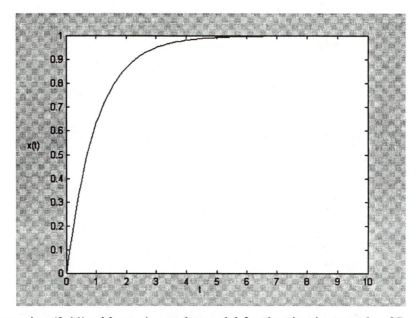

If equation (3.11) with $q = 1$ was the model for the circuit example of Problem 5 in Chapter 2, Figure 3.1 would show the voltage across the capacitor as a function of time. It starts at zero and in about five time units it settles at the value 1.

As another example, consider the differential equation

$$\frac{dx(t)}{dt} + x(t) = 3u(t)$$

where the interval of interest is $0 \le t \le 10$ and $x(0) = -2$. The solution will be made up of two parts, $x(t) = x_h(t) + x_p(t)$ where $x_h(t) = fe^{-t}$, which corresponds to the solution of equation $dx(t)/dt + x(t) = 0$; and $x_p(t) = a$. In this case $a = 3$ (plug $x_p(t)$ into the equation and set the right side equal to the left side). The complete solution has the form $x(t) = fe^{-t} + 3$. The value of f is then computed from the initial condition ($x(0) = -2 = f + 3, f = -5$), resulting in $x(t) = -5e^{-t} + 3$.

Second Order
Differential
Equations

With the experience gained from solving first order differential equations, let us now turn our attention to second order differential equations. We refer to them this way because of the presence of a *second* derivative of a function $x(t)$

.

An example is the differential equation:

$$\frac{d^2x(t)}{dt^2} = 1 \qquad\qquad (3.14)$$

with an interval of interest $0 \le t \le 10$. We can proceed in exactly the same manner as above and compute a solution by guessing. We ask: what is a function of time that when differentiated twice is equal to 1? Using the experience gained above, it would seem that a function like $x(t) = t^2$ would be a likely candidate. Differentiating this twice yields 2, so $x(t) = 1/2\ t^2$ would be just fine. But again, if we added anything to this function, which when differentiated twice becomes zero, it too would be a solution. Namely, $x(t) = 1/2\ t^2 + at + b$, is also a solution, where a and b are arbitrary constants. From this expression it is apparent that in order to identify the values of a and b and come up with a unique solution, we need to specify "two initial conditions." Suppose then we also require that the initial conditions $x'(0) = 2$ and $x(0) = 1$, where $x'(t)$ again denotes the derivative of $x(t)$ with respect to time, are also satisfied. This would force $b = 1$ and $a = 2$. So the solution of equation (3.14), which also satisfies the initial conditions, is given in equation (3.15) and plotted in Figure 3.3.

$$x(t) = \frac{1}{2}\ t^2 + 2t + 1 \qquad\qquad (3.15)$$

Equation (3.14) could be the model of the mechanical system shown in Figure 2.2, which is a mass with an external force equal to 1 acting on it. Expression (3.15) implies that the mass will start moving in the direction of the arrow and never stop! This seems to indicate a rather strange dynamic behavior. One could argue that this is because there is no friction present, but even if some were present, the result would have the same characteristics. We will return to this topic in Chapter 5, but for now let us say that such dynamic behavior in a physical system is undesirable.

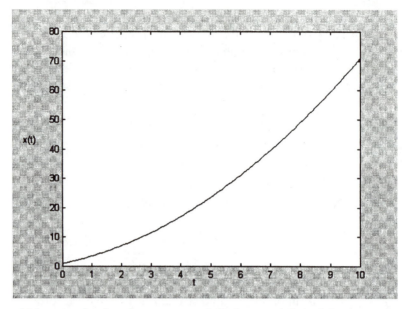

Figure 3.3
A Plot of the Solution to Equation (3.14)

We could have arrived at the same result had we employed the second method mentioned earlier, which involves integration. Again one can show that expression (3.15) is the unique solution to equation (3.14), but this is beyond the scope of our discussion,

Let us now consider the differential equation:

$$\frac{d^2x(t)}{dt^2} + \frac{dx(t)}{dt} = 0 \qquad (3.16)$$

where the interval of interest is $0 \leq t \leq 10$. In order to obtain a unique solution, we need to impose two initial conditions (on $x'(0)$ and $x(0)$). Suppose we now guess a solution of the form $x(t) = 1/2\ t^2 + at + b$, where a and b are constants. Plugging into equation (3.16), we have that the left side, which becomes $1 + t + a$, should be equal to the right side, which is zero. This cannot happen! Suppose that we try $x(t) = at + b$. Then we would have that a must be equal to 0. So it seems that $x(t) = b$ would work. However, it turns out that guessing functions of a different form will also work. The exponential function encountered above, when substituted, satisfies the equation. Specifically, let $x(t) = ae^{-t}$ where a is a constant. Then it is easily seen that since the second derivative of this function is ae^{-t} and the first derivative is $-ae^{-t}$, this function also solves the differential equation. Forming the sum $x(t) = ae^{-t} + b$ of these two expressions, we can verify that it too is a solution (differentiation is a "linear operator"). Let us also impose initial conditions: $x'(0) = 1$ and $x(0) = 2$.

This implies that $x'(0) = -a$, $x(0) = a + b$, which results in $a = -1$ and $b = 3$. The solution of differential equation (3.16), with the given initial conditions, is equal to:

$$x(t) = -e^{-t} + 3 \qquad (3.17)$$

Again one can prove, a fact beyond the scope of this book, that the solution just computed is unique. A plot of the solution is given in Figure 3.4.

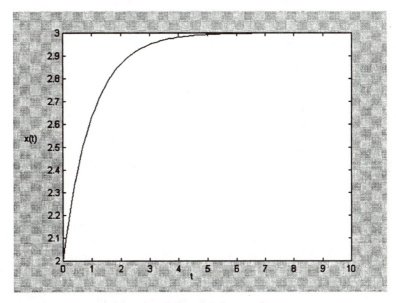

Suppose that we now consider the differential equation:

$$\frac{d^2x(t)}{dt^2} + 3\frac{dx(t)}{dt} + 2x(t) = 0 \qquad (3.18)$$

over the interval $0 \le t \le 10$, where in addition we have the initial conditions $x'(0) = 0$ and $x(0) = 1$. This would be the differential equation of the spring-mass-dashpot system depicted in Figure 2.4, where $m = 1$, $p = 3$, $k = 2$ and $f(t) = 0$. The initial position of the mass would be 1, and the initial velocity 0. Our solution strategy all along has been to "guess" at the form of the solution. The reader must agree that in view of the previous examples, it has not been a difficult task. However, even though from our earlier analysis we might speculate that an exponential form might work, trial and error will take a long time. In what follows we will suggest a way of identifying the appropriate exponential functions more directly.

We again associate a certain *polynomial* with differential equation (3.18). We replace the term $x(t)$ with the constant 1 the term $dx(t)/dt$ by the variable s and the term $d^2x(t)/dt^2$ by the variable s^2. Higher order derivative terms, had they appeared, would have been transformed in a similar fashion. This "transformation" takes terms involving time derivatives and replaces them with algebraic terms. In fact, it takes variables that involve "time" and replaces them with the s variable, which has nothing to do with "time." Such transformations are very useful in mathematics and engineering. For the moment, let us accept this transformation as given. We will provide more explanation in Section 3.3. This process generates the polynomial in s:

$$s^2 + 3s + 2 \qquad (3.19)$$

This is a polynomial in s of degree 2 and in general it has two roots (i.e., the *algebraic equation* $s^2 + 3s + 2 = 0$ has two solutions). We again call this the *characteristic polynomial* associated with the differential equation. The roots r_1 and r_2 of a general second-degree polynomial $as^2 + bs + c$, $(a \neq 0)$ are given by the quadratic formula, namely:

$$r_1 = \frac{-b + \sqrt{b^2 - 4ac}}{2a} \quad , \quad r_2 = \frac{-b - \sqrt{b^2 - 4ac}}{2a} \qquad (3.20)$$

The roots of the characteristic polynomial are also called the *poles* of the system and they can be *complex numbers*. We can actually distinguish three cases: these roots can be (1) real and unequal, (2) real and equal, or (3) two complex conjugate numbers. We will treat each case separately, as the form of the solution will be different in each case. In the discussion that follows, we assume that the coefficients a and b of the characteristic polynomial are *nonzero* real numbers.

REAL AND UNEQUAL ROOTS

For equation (3.18) the characteristic polynomial is equal to $s^2 + 3s + 2$, which has two real and unequal roots, $r_1 = -1$ and $r_2 = -2$. One can show that in this case, the appropriate exponential functions to try in solving differential equation (3.18) are: $x(t) = e^{r_1 t}$ and $x(t) = e^{r_2 t}$. For this example this implies that $x(t) = fe^{-t}$ and $x(t) = ge^{-2t}$ are solutions, where f and g are constants. We verify this by substituting $x(t) = fe^{-t}$ in (3.18):

$$fe^{-t} - 3fe^{-t} + 2fe^{-t} = 0$$

Similarly, if we substitute $x(t) = ge^{-2t}$ we have:

$$4ge^{-2t} - 6ge^{-2t} + 2ge^{-2t} = 0$$

Because of linearity we will also have that the sum $x(t) = fe^{-t} + ge^{-2t}$ satisfies the differential equation

$$fe^{-t} + 4ge^{-2t} - 3fe^{-t} - 6ge^{-2t} + 2fe^{-t} + 2ge^{-2t} = 0$$

The initial conditions will be used to specify the correct values for f and g. In particular, $x(0) = f + g = 1$ and $x'(0) = -f - 2g = 0$. These are two algebraic equations in two unknowns, which can easily be solved. If we add them, we have $-g = 1$, so $g = -1$. This implies that $f = 2$. The solution of differential equation (3.18), which in addition satisfies the given initial conditions, is:

$$x(t) = 2e^{-t} - e^{-2t} \qquad (3.21)$$

One can again show that this solution is unique. A plot of the solution is given in Figure 3.5. Earlier we pointed out that equation (3.18) could be the model of the system shown in Figure 2.4. If this is the case, it would be interesting to give a physical interpretation to the solution. The mass starts at position $x(0) = 1$, with zero velocity. The spring is compressed and will immediately begin to push the mass away from the wall. The mass comes to "rest" at position $x = 0$ in about seven units of time.

Figure 3.5
A Plot of the Solution to Equation (3.18)

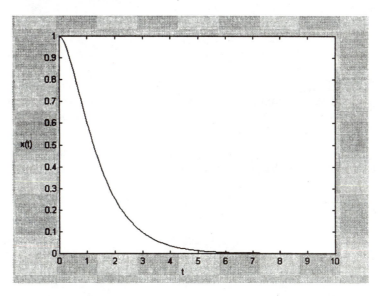

REAL AND EQUAL ROOTS

When the two roots of the characteristic polynomial are *equal* (i.e., $r_1 = r_2$) then the form of the two components of the solution will be $x(t) = fe^{r_1 t}$ and $x(t) = gte^{r_1 t}$. Note that the second component is an exponential multiplied by the function t. One can easily verify that the solution of equation $d^2x(t)/dt^2 +$

$2dx(t)/dt + 1x(t) = 0$ is: $x(t) = fe^{-t} + gte^{-t}$. This is because the characteristic polynomial $(s^2 + 2s + 1)$ has two real and equal roots at -1. As before, one can compute the values of f and g from the initial conditions.

COMPLEX CONJUGATE ROOTS

Let us now turn our attention to a situation in which the roots of the second-degree characteristic polynomial that corresponds to the differential equation are complex conjugate numbers. In particular, consider the differential equation:

$$\frac{d^2x(t)}{dt^2} + \frac{dx(t)}{dt} + 2.5x(t) = 0 \qquad (3.22)$$

where the initial conditions are $x(0) = -1$, $x'(0) = 1$. The corresponding characteristic polynomial in s is:

$$s^2 + 1s + 2.5$$

The two roots are given by the quadratic formula and can easily be computed as:

$$r_1 = \frac{-1 + \sqrt{1 - 10}}{2} = -.5 + j1.5, \qquad r_2 = \frac{-1 - \sqrt{1 - 10}}{2} = -.5 - j1.5$$

where $j = \sqrt{-1}$. These are complex numbers with real part equal to -.5 and imaginary part equal to ±1.5 (they are complex conjugates). The form of the functions that should be tried as solutions is somewhat different than before. However, one can easily check that the two functions $ae^{-.5t}\cos(1.5t)$, and $be^{-.5t}\sin(1.5t)$ both solve the differential equation. The real part of the root becomes the constant that multiples t in the exponent of the exponential term and the absolute value of the imaginary part multiplies t in the sin and cos terms. Let us verify that $e^{-.5t}\cos(1.5t)$ is a solution:

$$\frac{dae^{-.5t}\cos(1.5t)}{dt} = -.5ae^{-.5t}\cos(1.5t) - 1.5ae^{-.5t}\sin(1.5t)$$

and

$$\frac{d^2ae^{-.5t}\cos(1.5t)}{dt^2} = -2ae^{-.5t}\cos(1.5t) + 1.5ae^{-.5t}\sin(1.5t)$$

Substituting these expressions in the differential equation we have:

$$-2ae^{-.5t}\cos(1.5t) + 1.5ae^{-.5t}\sin(1.5t)$$
$$- .5ae^{-.5t}\cos(1.5t) - 1.5ae^{-.5t}\sin(1.5t) + 2.5ae^{-.5t}\cos(1.5t) = 0$$

In a similar manner, we can also verify that $be^{-.5t}\sin(1.5t)$ is a solution as well and since the equation is linear, their sum will be also The general solution is therefore given by:

$$x(t) = ae^{-.5t}\cos(1.5t) + be^{-.5t}\sin(1.5t)$$

where a and b are constants, whose value is determined form the initial conditions. In particular, $x(0) = a = -1$ and since

$$\frac{d(ae^{-.5t}\cos(1.5t) + be^{-.5t}\sin(1.5t))}{dt} =$$
$$(-.5a + 1.5b)e^{-.5t}\cos(1.5t) - (1.5a + .5b)e^{-.5t}\sin(1.5t)$$

we must have $x'(0) = (-.5a + 1.5b) = 1$. Solving these two algebraic equations in a and b we have $a = -1$ and $b = 1/3$. So the solution of differential equation (3.22), with the given initial conditions, is:

$$x(t) = -e^{-.5t}\cos(1.5t) + \frac{1}{3} e^{-.5t}\sin(1.5t) \tag{3.23}$$

One can again show that the computed solution is unique. A plot of the solution is given in Figure 3.6.

Figure 3.6
A Plot of the Solution to Equation (3.22)

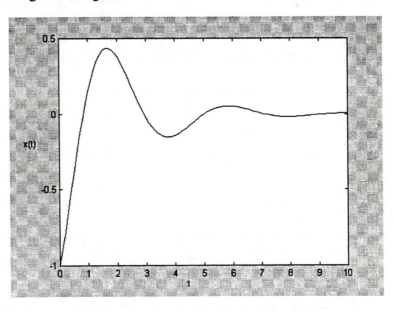

We have just seen how to compute solutions of second order linear differential equations with constant coefficients and how the form of the solution depends on the roots of the associated characteristic polynomial. Our main focus was on the situation in which the right side of the equation is equal to the zero function. However, just as in the case of first order differential equations, the right side need not be equal to the zero function. We could have it be some other function of time. In this book we will only consider the case in which this function is equal to a constant (i.e., the right side is a step function). Specifically, suppose that we have the differential equation:

$$\frac{d^2x(t)}{dt^2} + 3\frac{dx(t)}{dt} + 2x(t) = 2u(t) \qquad (3.24)$$

where the function $u(t)$, the *input*, is again the *unit step* function:

$$u(t) = \begin{cases} 0 & t < 0 \\ 1 & t \geq 0 \end{cases} \qquad (3.25)$$

the interval of interest is $0 \leq t \leq \infty$ and in addition, we have the initial conditions $x'(0) = 0$ and $x(0) = 0$. Equation (3.24) could be the model of the circuit shown in Figure 2.8, where $x(t)$ is the capacitor voltage, $R = 3$, $L = 1$, $C = 1/2$, and $u(t)$ is the battery voltage. The initial capacitor voltage is zero and so is the initial current that flows in the circuit. One can show that for this, as well as for the general second order linear differential equation with constant coefficients, the solution must now be modified by adding an additional part, exactly as in the case of first order differential equations. The solution will thus have two parts:

- the first part is the one computed for the case when the right side is equal to the zero function (as above), which has the form $x_h(t) = fe^{-t} + ge^{-2t}$; the homogeneous solution.

- the second part for a step input will be a constant $x_p(t) = a$; the particular solution.

So the complete solution will be of the form

$$x(t) = x_h(t) + x_p(t) = fe^{-t} + ge^{-2t} + a$$

The value of the constant a can again be obtained by substituting $x_p(t) = a$ in the equation. For this case, we have for any nonnegative t that $2a = 2$. So $a = 1$, $x_p(t) = 1$ and the complete solution will be $x(t) = fe^{-t} + ge^{-2t} + 1$. The constants f and g can now be computed by using the initial conditions. In particular, $x(0) = 0 = f + g + 1$, and $x'(0) = 0 = -f - 2g$. Solving these two equations for f and g, we obtain $f = -2$ and $g = 1$. The complete solution of (3.20) is given below and plotted in Figure 3.7 for $0 \le t \le 10$.

$$x(t) = -2e^{-t} + e^{-2t} + 1$$

Figure 3.7
*A Plot of the
Step Response
for Equation
(3.24)*

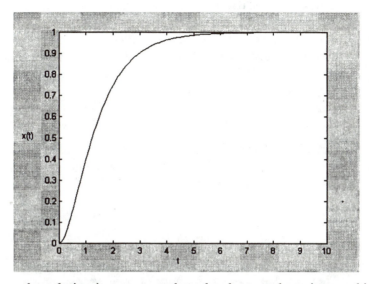

Here, too, the solution just computed can be shown to be unique and is referred to as the *step response*. If equation (3.24) models the circuit of Figure 2.8, the physical interpretation of this graph is that the capacitor voltage starts at 0 and steadily increases to the value 1.

Equation (3.24) has a characteristic polynomial with two real and unequal roots. The solution method just presented applies not only to this specific example, but to the general second order linear differential equation with nonzero constant coefficients and a "step" input. The "homogeneous part" of the solution would be different in each case, as dictated by the roots of the corresponding characteristic polynomial.

In this introductory exposition to differential equations we confined our discussion to first and second order linear differential equations with constant coefficients, where the input $u(t)$ is either the function zero or the step input. Here we will not present the theory of higher order linear differential equations with constant coefficients or deal with the general case when $u(t)$ is some other function of time. This will be done in future differential equations courses, where the reader will find that the basic solution methods presented here continue to apply (see Boyce and Diprima, 1977).

What we have done thus far is show how to obtain *analytical* solutions (i.e., derive formulas for the solutions) of differential equations. Our next topic will be a discussion on how to obtain approximate solutions to simple differential equations *numerically*. One can think of this task as asking a digital computer, without symbolic mathematics capabilities, to compute the solution of a differential equation. Computers can handle finite sets (albeit huge) of numbers, which can be represented by a finite number of "bits." A computer cannot, for example, save all the real numbers in the interval from zero to one, as there is an uncountably infinite many of them. When we numerically solve a differential equation by digital computer we will call the process *simulation* (i.e., we perform a digital computer simulation of a differential equation). We will focus on one method, the so called *Euler* method, but many more exist. In the next section we will present the theory by considering specific simple examples. In the next chapter we will discuss a specific software package, MATLAB/SIMULINK, which allows us to perform simulations in a very efficient and user-friendly manner. There we will see how to simulate not only second order systems but more complicated ones as well.

Before we embark on the study of numerical solutions to differential equations, it is appropriate to say something more about the term *linear,* which was used to describe the differential equations under study. Consider the equations:

$$\frac{dx(t)}{dt} + bx(t) = 0 \qquad \frac{d^2x(t)}{dt^2} + a\frac{dx(t)}{dt} + bx(t) = 0$$

They are called *linear* because of their special property that if $x_1(t)$ and $x_2(t)$ are both solutions to any one of them, then $x_1(t) + x_2(t)$ is also a solution and so are $ax_1(t)$ and $bx_2(t)$ where a and b are constants. We have used this property repeatedly in this section. This would not be true for the equation

$$\frac{dx(t)}{dt} + bx^2(t) = 0$$

because of the presence of the term $x^2(t)$. This equation and many others like it are called *nonlinear differential equations*. The study of solutions to nonlinear differential equations is more complex and will not be addressed here. We will, however, encounter nonlinear behavior in Chapter 7 while developing models and control strategies for a computer-controlled model car.

3.2 Numerical Solutions of Differential Equations

In Section 3.1 we saw how to provide analytical solutions to simple differential equations. In other words, we gave formulas for the solution. Currently, there are software packages commercially available (e.g., MACSYMA, MAPLE or MATHEMATICA) that can perform symbolic computations and generate analytical solutions to such simple differential equations. However, one can also choose to compute the solution numerically. Specifically, rather than obtaining a formula for the solution $x(t)$ one can obtain a finite sequence $x_a(t_i)$, $1 \leq i \leq N$, of numbers such that $x_a(t_i) \approx x(t_i)$, for all $1 \leq i \leq N$. This is particularly useful in the solution of differential equations by digital computer, as well as in plotting a solution. This is precisely what we did when we generated the plots of solutions in Section 3.1. Clearly, this is not in as convenient a form as having an analytical expression for the solution. In future studies the reader will see that in many cases this is the only way in which solutions can be computed. There are many different ways of computing numerical solutions to differential equations. Here we have chosen to briefly discuss the so-called *Euler* method because of its simplicity. Let us present it by example. In Section 3.1 we considered the following differential equation:

$$\frac{dx(t)}{dt} = 1 \qquad (3.26)$$

Let us assume that the initial condition is $x(0) = 0$ and that we are interested in the solution in the interval $0 = t = 10$. We already know that the solution is $x(t) = t$. One of the methods suggested for solving this differential equation, was via integration rather than "guessing." Specifically, we wrote:

$$\int_0^v x'(t)dt = \int_0^v 1dt \qquad (3.27)$$

which implies that $x(v) - x(0) = v$, and since $x(0) = 0$, $x(v) = v$ (or, changing the independent variable to t, $x(t) = t$) is the exact analytical expression.

Suppose that we proceeded in a different way. We divided the interval $0 \leq t \leq 10$ into subintervals of length $h = .1$ (generating 100 intervals). Define:

$$x(k\,h) = \int_0^{kh} 1dt \qquad \text{where } 1 \leq k \leq 100 \qquad (3.28)$$

Then using properties of definite integrals we can write:

$$x(k\,h) = \int_0^{(k-1)h} 1dt \ + \ \int_{(k-1)h}^{kh} 1dt \ = x((k-1)h) + \int_{(k-1)h}^{kh} 1dt \qquad (3.29)$$

Now this expression, which is referred to as a *difference equation*, gives us an idea of how to compute the solution. This equation says that the value of the solution at the kth time step $x(k\,h)$ is equal to the value of the solution at the $(k$

- 1)th step $x((k - 1)h)$, plus something more, which in this case we can compute and it is equal to h. So we can rewrite equation (3.29) as:

$$x(kh) = x((k - 1)h) + h \qquad (3.30)$$

where k takes values in the range $1 \leq k \leq 100$. This allows us to write:

For $k = 1$

$$x(h) = x(0) + h = h = .1$$

For $k = 2$

$$x(2h) = x(h) + h = h + h = 2h = .2$$

For $k = 3$

$$x(3h) = x(2h) + h = 3h = .3$$

We can proceed recursively in this manner and compute all $x(kh)$ for k in the range $1 \leq k \leq 100$, ending with $100h = 10$. In this case these 100 values *match exactly* the values of the solution at the corresponding points of time. However, we do not have the values of the solution at in-between points, say at $t = .26$ or $t = 7.543$. This method seemed to have worked quite well in this example, even though it does not give us the solution in an analytical form ($x(t) = t$). Let us investigate what would happen if we used it in the solution of the equation

$$\frac{dx(t)}{dt} = -x(t) \qquad (3.31)$$

where we also assume that the initial condition is $x(0) = 1$. We know from our analysis in Section 3.1 that the exact analytical solution is given by $x(t) = e^{-t}$. However, assume that we did not know this and let us use the integration method presented above to compute the solution at time instants kh where k is a positive integer and $h = .1$. Integrating both sides of (3.31) we have:

$$\int_0^v x'(t)\,dt = -\int_0^v x(t)\,dt$$

The left side is equal to $x(t) - x(0)$. This implies that for any v, we must have:

$$x(v) - x(0) = - \int_0^v x(t)\,dt \qquad\qquad (3.32)$$

Specifically, for $v = kh$ and $v = (k - 1)h$ we have:

$$x(kh) - x(0) = - \int_0^{kh} x(t)\,dt \quad \text{and} \quad x((k - 1)h) - x(0) = - \int_0^{(k-1)h} x(t)\,dt \qquad (3.33)$$

We cannot compute the right side integral of these two expressions, since we assumed we do not know $x(t)$. However, we can use a basic property of integrals to write:

$$\int_0^{kh} x(t)\,dt \;=\; \int_0^{(k-1)h} x(t)\,dt \;+\; \int_{(k-1)h}^{kh} x(t)\,dt \qquad (3.34)$$

Using (3.33), equation (3.34) becomes:

$$-x(kh) + x(0) \;=\; - x((k - 1)h) + x(0) + \int_{(k-1)h}^{kh} x(t)\,dt \;\;,$$

$$x(kh) = x((k - 1)h) - \int_{(k-1)h}^{kh} x(t)\,dt \qquad (3.35)$$

Notice again that this expression is a difference equation, as the kth value of x is related to the $(k - 1)$st value of x. However, unlike the previous example where we were able to compute the integral, here we have assumed that we do not know the expression for $x(t)$. Since we cannot compute the integral in the right side, we will approximate it. First, we should point out that because of this approximation, we will no longer obtain an exact solution. Second, this approximation can be done in many ways. Perhaps the simplest one is the Euler method. It is derived from the realization that a definite integral represents area under the curve. Specifically, let $f(t)$ be any function, then:

$$\int_a^b f(t)\,dt$$

is the area under the curve $f(t)$ corresponding to the interval (a,b). Let us apply this to the problem at hand. Let $x(t)$ be the function in question and let $a = (k - 1)h$ and $b = kh$. We can approximate the area under the curve $x(t)$ by a rectangle (see shaded area in Figure 3.8).

Figure 3.8
Approximation of the Area Under a Curve

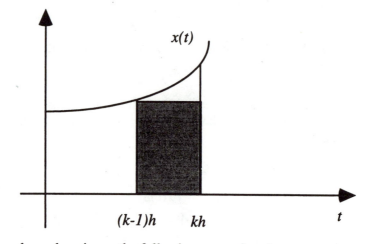

What we have done is use the following approximation:

$$\int_{(k-1)h}^{kh} x(t)dt \approx x((k - 1)h)h \qquad (3.36)$$

Notice that by using this approximation, we have been able to replace the integral we could not compute by an approximation that we can use, in the sense that we can relate it to the function values at the subinterval endpoints. Specifically, what we now do is replace the integral in (3.35) and write:

$$x_a(kh) = x_a((k - 1)h) - x_a((k - 1)h)h = x_a((k - 1)h)(1 - h) \qquad (3.37)$$

It is now proper to replace x by some other variable x_a, because we are now dealing with an "approximation." This expression is again a difference equation that can be solved recursively. We can start with $x_a(0) = x(0) = 1$, and then compute the values $x_a(kh)$ recursively using (3.37). For this example, where we assume $h = .1$, we will have (the symbol "$\times$" denotes multiplication):

For $k = 1$

$$x_a(h) = x_a(.1) = x_a(0)(1 - h) = 1 \times .9 = .9$$

For $k = 2$

$$x_a(2h) = x_a(.2) = x_a(h)(1 - h) = .9 \times .9 = .81$$

For $k = 3$

$$x_a(3h) = x_a(.3) = x_a(2h)(1 - h) = .81 \times .9 = .729$$

For $k = 4$

$$x_a(4h) = x_a(.4) = x_a(3h)(1 - h) = .729 \times .9 = .6561$$

and in this way we can compute the approximate solution for all k. We recognize that there is a pattern here:

$$x_a(h) = (1 - h)$$
$$x_a(2h) = (1 - h)^2$$
$$x_a(3h) = (1 - h)^3$$
$$x_a(4h) = (1 - h)^4$$

$$\cdot$$
$$\cdot$$

and in general $x_a(kh) = (1 - h)^k$ for any nonnegative k. This method has allowed us to obtain an approximate solution to our differential equation. It is natural to want to know how close this approximation is to the exact solution. Table 3.1 compares the two for the first few time instants. We actually report approximate solutions from two cases, one for $h = .1$ and another for $h = .01$ (see discussion below).

t	$x(t)$	$x_a(t),\ h = .1$	$x_a(t),\ h = .01$
0	1	1	1
.1	0.904837418036	.9	0.904382075009
.2	0.818730753078	.81	0.817906937597
.3	0.740818220682	.729	0.739700373388
.4	0.670320046036	.6561	0.668971758570

Table 3.1 Comparison of Exact and Approximate Solutions

One can see that the entries in columns two and three are close. In fact, we could easily see how we could make the approximation even better by reducing the error introduced in the rectangular approximation of the area under a curve. The smaller the width of the rectangle, the better the approximation (see Figure 3.8). Suppose now that the interval is chosen to be $h = .01$. In the above analysis we provided expressions for an arbitrary h so $x_a(.1) = x_a(10h) = (1 - .01)^{10} = 0.904382075009$, $x_a(.2) = x_a(20h) = (1 - .01)^{20} = 0.817906937597$, $x_a(.3) = x_a(30h) = (1 - .01)^{30} = 0.739700373388$, $x_a(.4) = x_a(40h) = (1 - .01)^{40} = 0.668971758570$, and so on. This information is depicted in column four of Table 3.1. In Figure 3.9 the solutions just computed are plotted in the range $0 \leq t \leq 10$.

Figure 3.9
Plot of Exact and Approximate Solutions of Equation (3.31)

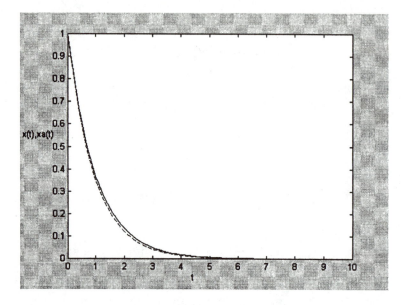

In the plot the solid line is the exact solution, the dashed line is the approximate solution computed with $h = .1$. The approximate solution computed with $h = .01$ falls right on top of the solid line and can hardly be distinguished from the exact solution.

There are several other methods that can be used to compute approximate solutions to differential equations (e.g., Runge-Kutta method). All are based on the idea presented above, of approximating the area under a curve in some way that results in a difference equation involving the approximate solution at the "sampling instants." The resulting difference equation can then be solved recursively.

We presented the method in terms of a specific example, but we should mention that it can be used for large classes of equations. The next example shows how it can be used for the computation of the approximate solution of a differential equation, with a nonzero input (i.e., a nonzero right side). Let us compute numerically the solution of the differential equation (in the interval $0 \leq t \leq \infty$):

$$\frac{dx(t)}{dt} + x(t) = u(t) \qquad (3.38)$$

where the initial condition is $x(0) = 0$ and $u(t)$ is the unit step input (see (3.25)). We know from our earlier analysis that the exact analytical solution is given by $x(t) = 1 - e^{-t}$ for $t \geq 0$. Let's use the numerical integration method considered above to compute the solution at time instants kh, where k is a positive integer and $h = .1$. Integrating both sides of (3.38) we have:

$$\int_0^v x'(t)dt + \int_0^v x(t)dt = \int_0^v u(t)dt$$

Since $x(0) = 0$, this becomes:

$$x(v) + \int_0^v x(t)dt = \int_0^v u(t)dt$$

For specific values of $v = (k - 1)h$ and $v = kh$ we have:

$$x(kh) + \int_0^{kh} x(t)dt = \int_0^{kh} u(t)dt \;, \quad x((k-1)h) + \int_0^{(k-1)h} x(t)dt = \int_0^{(k-1)h} u(t)dt \qquad (3.39)$$

Subtracting the second equation in (3.39) from the first results in:

$$x(kh) - x((k-1)h) + \int_0^{kh} x(t)dt - \int_0^{(k-1)h} x(t)dt = \int_0^{kh} u(t)dt - \int_0^{(k-1)h} u(t)dt$$

and using the basic property of integrals:

$$\int_0^{kh} x(t)dt = \int_0^{(k-1)h} x(t)dt + \int_{(k-1)h}^{kh} x(t)dt$$

for both the $x(t)$ and $u(t)$ integrals gives:

$$x(kh) = x((k-1)h) - \int_{(k-1)h}^{kh} x(t)dt \quad + \quad \int_{(k-1)h}^{kh} u(t)dt \qquad (3.40)$$

Since $u(t)$ is a step function the second integral on the right side is just equal to h. We cannot compute the first integral on the right side and so we will approximate it in the same manner as in the previous example:

$$\int_{(k-1)h}^{kh} x(t)dt \quad \approx x((k-1)h)h$$

Using this approximation we are able to replace the integral we cannot compute by an approximate expression in terms of the values of the function at the "sampling instants." Since we will be computing approximate solutions we replace the variable "x" with the variable "x_a" and use the above approximation for the integral term to write:

$$x_a(kh) = x_a((k-1)h) - x_a((k-1)h)h + h = x_a((k-1)h)(1-h) + h \quad (3.41)$$

This expression is again a difference equation that can be solved recursively. We can start with $x_a(0) = x(0) = 0$ and then compute the values $x_a(kh)$ recursively using (3.41). For this example, where we assume $h = .1$, we have:

For $k = 1$ $\qquad\qquad x_a(h) = x_a(.1) = x_a(0).9 + .1 = .1$

For $k = 2$ $\qquad\qquad x_a(2h) = x_a(.2) = x_a(h).9 + .1 = .19$

For $k = 3$ $\qquad\qquad x_a(3h) = x_a(.3) = x_a(2h).9 + .1 = .271$

For $k = 4$ $\qquad\qquad x_a(4h) = x_a(.4) = x_a(3h).9 + .1 = .3439$

and in this way we can compute the approximate solution for all k. As in the previous example we can recognize that there is a pattern here:

$$x_a(h) = h$$
$$x_a(2h) = (1 + .9)h$$
$$x_a(3h) = (1 + .9 + (.9)^2)h$$
$$x_a(4h) = (1 + .9 + (.9)^2 + (.9)^3)h$$

and for an arbitrary k we have:

$$x_a(kh) = (1 + .9 + (.9)^2 + (.9)^3 + \ldots + (.9)^{(k-1)})h$$

Table 3.2 compares the actual solution values at the sampling instants kh with the approximate solution values at the same time instants.

t	$x(t)$	$x_a(t)$
0	0	0
.1	.095162582	.1
.2	.181269247	.19
.3	.259181779	.271
.4	.329679954	.3439

Table 3.2 Comparison of Exact and Approximate Solutions

We can plot the actual solution and the approximation in order to compare the results. Even for a step size $h = .1$ the difference is quite small. We could have obtained a better approximation if we had used a smaller h. In Figure 3.10 the actual solution is the solid line and the approximation is depicted as a dashed line. Since the right side of the differential equation (input) was a unit step function, the solution is the step response.

Figure 3.10
Plots of Exact and Approximate Solutions for Equation (3.38)

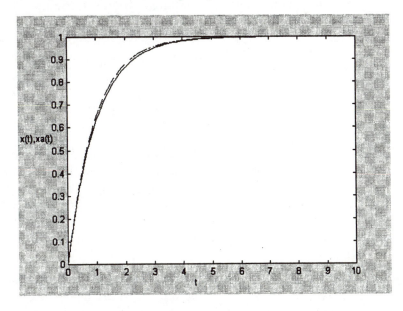

The plot shows the results of a numerical computation of the (approximate) solution to a first order differential equation. Now that we know how numerical solutions are computed (we know the background theory) it would be great if there were a more convenient method of carrying out the procedure. In fact, we will see in the next chapter how this can be done. User-friendly software packages have been developed and are in use. We will become familiar with one called SIMULINK, which is part of a larger package called MATLAB. We will see that obtaining step responses is particularly easy.

We have just discussed numerical techniques for solving first order linear differential equations. These same techniques can be generalized and applied to second order differential equations, more complicated linear differential equations and nonlinear differential equations. We focused on the Euler method because of its simplicity. More accurate methods exist (e.g., Runge-Kutta), descriptions of which can be found in Boyce and DiPrima (1977).

3.3　Transfer Function System Models

In Chapter 2 we saw how differential equations can be used to describe, or model, the operation of dynamic systems. These were linear differential equations and so the system models are called *linear*. Since these differential equations have constant coefficients, the system models are called *time invariant*. In Sections 3.1 and 3.2 we presented methods for obtaining exact (analytical) solutions and approximate solutions, respectively. We have stressed repeatedly that these dynamic representations can be thought of as input-output relationships, with the input being the right side of the equation and the output being the solution. In this section, we will carry this input-output representation one step further and introduce the concept of a system *transfer function*. Suppose that a system is modeled by the following second order linear differential equation with constant coefficients:

$$4\frac{d^2x(t)}{dt^2} + 2\frac{dx(t)}{dt} + 6x(t) = f(t) \qquad (3.42)$$

For a specific input $f(t)$ and initial conditions $x(0) = 0$ and $x'(0) = 0$, one can in principle compute the solution $x(t)$ (output) for all time $t = 0$, either numerically or analytically. The system takes the specific input $f(t)$ and turns it into a particular output $x(t)$. If the initial conditions were not zero, then the output (solution) would be different. In order to be consistent in comparing these solutions for different systems, let us assume that the initial conditions are zero.

Earlier in this chapter we talked about a transformation that takes the derivative terms in a differential equation and replaces them with algebraic terms involving the variable s. In this way we generated the characteristic polynomial associated with a particular linear time invariant differential equation and proceeded to compute solutions. That process was "part" of the so-called *Laplace Transformation*. Actually, this Laplace Transformation can be used to develop a dynamic model for a system as in (3.42), which is an *algebraic relationship* and not a differential equation. This is done in the following straightforward manner: On the left side of the equation, the function $x(t)$ is replaced by a new function $X(s)$ (called its *Laplace Transform*), the function $dx(t)/dt$ is replaced by $sX(s)$ (i.e., s times $X(s)$) and the function $d^2x(t)/dt^2$ is replaced by $s^2X(s)$ (i.e., s^2 times $X(s)$). On the right side of the equation (3.42) a similar "replacement" is carried out. The function $f(t)$ is replaced by the function $F(s)$, its Laplace Transform. In this example there are no time derivatives of $f(t)$ that appear on the right side, but had there been, we would have dealt with them in exactly the same manner (replacing the function $df(t)/dt$ by the function $sF(s)$, etc.). Actually there is a formula that is used to compute $X(s)$ from $x(t)$, and one for computing $x(t)$ from $X(s)$ ($x(t)$ is called the inverse Laplace Transform of $X(s)$). At the end of this section there is a more detailed explanation of this procedure. After completing this transformation process we have the following algebraic equation:

$$4s^2X(s) + 2sX(s) + 6X(s) = F(s) \qquad (3.43)$$

This algebraic equation relates the transformed variables $X(s)$ and $F(s)$ and it too can be thought of as a dynamic description of the same system. The "information" carried by either equation (3.42) or (3.43) is precisely the same. However, one can immediately see that this algebraic equation can now be very easily "solved" for $X(s)$:

$$(4s^2 + 2s + 6)X(s) = F(s)$$

Divide both sides by $(4s^2 + 2s + 6)$ and obtain:

$$X(s) = \frac{1}{4s^2 + 2s + 6} \; F(s) \qquad (3.44)$$

This is a much simpler relationship between the "transformed" input $F(s)$ and the "transformed" output $X(s)$ for this system. This expression simply says that $X(s)$ is just the product of the rational function $G(s) = 1/(4s^2 + 2s + 6)$, with the function $F(s)$. It says that $G(s)$ "transfers" the input $F(s)$ to the output $X(s)$. For this reason it is called the *Transfer Function* of the system. In this manner, transfer functions can be used to represent dynamic models for systems. These are the models that will be used in the next chapter to represent systems when we solve differential equations *numerically* by digital computer. Note that the denominator of the transfer function is the characteristic polynomial of the differential equation; consequently, this denominator is also called the characteristic polynomial.

Transfer functions are also used to compute analytical solutions of differential equations. This is quite evident from our discussion above, as equation (3.44) provides an expression for $X(s)$. The procedure can be summarized as follows: (1) the differential equation is "transformed" to an algebraic equation using the Laplace Transformation, (2) the Laplace Transform $X(s)$ of $x(t)$ is then computed and (3) the Inverse Laplace Transform is used to compute $x(t)$ from $X(s)$. The transformation of equation (3.42) to equation (3.43) was done here in a "mechanical" fashion. In what follows, we will present an example and show how this is done in a mathematically rigorous manner.

For the system in (3.44) the numerator polynomial is equal to unity because there are no terms on the right side of differential equation (3.42) that involve derivatives of $f(t)$. If there were such terms present, the numerator would also be a polynomial in s. Specifically, suppose that the system was described by the differential equation:

$$4\frac{d^2x(t)}{dt^2} + 2\frac{dx(t)}{dt} + 6x(t) = 3\frac{df(t)}{dt} + f(t) \qquad (3.45)$$

Then the transfer function that corresponds to this system is:

$$G(s) = \frac{3s + 1}{4s^2 + 2s + 6}$$

In the above discussion we began with a differential equation and developed an equivalent dynamic system representation using the Laplace Transformation. For the more advanced reader, we would like to provide some more detail regarding this process. The Laplace Transform $X(s)$ of some function $x(t)$ is defined by the following improper integral:

$$X(s) = \int_0^\infty x(t)e^{-st}\,dt \qquad (3.46)$$

where the variable s is considered to be a complex number. The transformation takes a function $x(t)$ and turns it into the new function $X(s)$ of the complex variable s. There is a very well-developed theory of this Laplace Transformation that is beyond the scope of this book (see Bobrow, 1987, and Ogata, 1997). Here we will confine our discussion to some basic facts relevant to our investigation. Specifically, one can show that there is a one-to-one relationship between $x(t)$ and $X(s)$. In other words, for every $x(t)$ there is a unique $X(s)$ that corresponds to it. There is also a formula involving integrals for $x(t)$ in terms of $X(s)$ that is called the Inverse Laplace Transformation. So knowing $X(s)$ is equivalent to knowing $x(t)$ and vice versa. Tables can be found in many books (e.g., Bobrow, 1987, and Ogata, 1997) that list the pairs $(x(t),\ X(s))$. Let us use the formula in (3.46), to compute the Laplace Transform for the unit step function $u(t)$ encountered above. We have:

$$U(s) = \int_0^\infty u(t)\, e^{-st}\, dt \ = \ \int_0^\infty 1 e^{-st}\, dt \ = -\frac{1}{s}\ e^{-st}\Big|_0^\infty \qquad (3.47)$$

Since this is an improper integral, caution must be used when computing it. The variable s can take many values. For some, this integral will have a finite value, but for others it will not. In particular, suppose that $s = 1$. When we evaluate (3.47) at the high limit, we will have: $\lim_{t \to \infty} e^{-t} = 0$. In fact, this will be true for any complex s that has a positive real part. On the other hand, if $s = -1$, then this limit "blows up." The range of values of s in the complex plane, for which this integral has a finite value, is called the "region of convergence." For any s in the region of convergence, the value of the integral in (3.47) is $1/s$. Therefore, we say that the Laplace Transform of the unit step function is:

$$U(s) = \frac{1}{s}$$

As a second example, consider the exponential function $x(t) = e^{-t}$. From (3.46) we have that its Laplace Transform is:

$$X(s) = \int_0^\infty x(t) e^{-st}\, dt \ = \ \int_0^\infty e^{-t} e^{-st}\, dt \ = \frac{-1}{s+1} e^{-(s+1)t}\Big|_0^\infty \qquad (3.48)$$

For values of s in the region of convergence of this integral (any complex s with real part greater than -1) one can easily see that $X(s) = 1/(s + 1)$. In both these examples and for many other important functions $x(t)$, the corresponding $X(s)$ is a rational function in s (i.e., a ratio of polynomials in s).

The beautiful aspect of the Laplace Transformation is that it takes differential equations and turns them into "equivalent" algebraic equations, which are much simpler dynamic descriptions for systems. This also facilitates their solution. Consider the differential equation:

$$\frac{dx(t)}{dt} + x(t) = u(t) \tag{3.49}$$

which is the model of some system and where $x(0) = 0$ and $u(t)$ is a unit step function. We already know that the Laplace Transform of $u(t)$ is $1/s$. Assume that $x(t)$ and $dx(t)/dt$ are functions that have Laplace Transforms and that the Laplace Transform of $x(t)$ is denoted by $X(s)$. Equation (3.49) states that the function on the left side is equal to that on the right side. This means that the Laplace Transforms of both sides should be the same:

$$\int_0^\infty (\frac{dx(t)}{dt} + x(t))e^{-st}\, dt \quad = \quad \int_0^\infty u(t)e^{-st}\, dt \tag{3.50}$$

Using the linearity property of integrals, this implies that:

$$\int_0^\infty \frac{dx(t)}{dt}e^{-st}\, dt \quad + \quad \int_0^\infty x(t)e^{-st}\, dt \quad = \quad \int_0^\infty u(t)e^{-st}\, dt$$

We immediately recognize that the right side is equal to $U(s) = 1/s$, and also notice that the second integral on the left side is just $X(s)$, the Laplace Transform of $x(t)$. Let us focus attention on the first integral of the left side and use the formula for "integration by parts." We have:

$$\int_0^\infty \frac{dx(t)}{dt}e^{-st}\, dt = x(t)e^{-st}\Big|_0^\infty - \int_0^\infty (-s)x(t)e^{-st}\, dt$$

For appropriate values of s that make $\lim_{t \to \infty} x(t)e^{-st} = 0$ and since $x(0) = 0$, the right side is equal to:

$$s \int_0^\infty x(t)e^{-st}\, dt \quad = sX(s)$$

Therefore, the Laplace Transform of $dx(t)/dt$ is just $sX(s)$. Returning to equation (3.50) we can write:

$$sX(s) + X(s) = U(s)$$

This is an algebraic expression in $X(s)$ that provides an equivalent dynamic characterization for the system in terms of the transfer function $1/(s + 1)$. Furthermore, this procedure allows us to solve for $X(s)$. Namely,

$$X(s) = \frac{1}{s+1} \quad U(s) = \frac{1}{s+1} \frac{1}{s} = \frac{1}{s} + \frac{-1}{s+1}$$

The corresponding $x(t)$ can now be computed from this $X(s)$, since one can recognize that this is the Laplace Transform of the function $f(t) = 1 - e^{-t}$, $t \geq 0$. Here we rely on the one-to-one property of the Laplace Transform.

The Laplace Transformation has allowed us to generate a transfer function model for a dynamic system. It has also allowed us to solve the corresponding differential equation. This methodology can be easily generalized and applied to a large class of systems. However, we should keep in mind that it works only when we have systems with dynamic models that are linear time differential equations with constant coefficients. We will use transfer function descriptions in all the remaining chapters–in Chapter 4 when we discuss digital computer simulations, in Chapter 5 when we introduce the notion of stability, in Chapter 6 when we discuss feedback configurations and in Chapter 7 when we develop a system model using experiments.

3.4   Exercises

Problem 1   Compute the solution of the following differential equation:

$$\frac{dx(t)}{dt} + 3x(t) = 0$$

in the interval $t \geq 0$, with initial condition $x(0) = -2$.

Problem 2   Compute the solution of the differential equation in Problem 1 numerically using Euler's method with $h = .1$ and $h = .01$. Generate a table showing the approximate values as well as the actual values at the first six sampling instants.

Problem 3   Compute the solution of the following differential equation:

$$\frac{dx(t)}{dt} - 2x(t) = 0$$

in the interval $t \geq 0$, with initial condition $x(0) = 3$.

Problem 4   Compute the solution of the following differential equation:

$$\frac{dx(t)}{dt} + 4x(t) = u(t)$$

in the interval $t \geq 0$, where $u(t)$ is the unit step function and $x(0) = 1$.

Problem 5   Compute the solution of the differential equation in Problem 4 numerically using Euler's method with $h = .1$ and $h = .01$. Generate a table showing the approximate values as well as the actual values at the first six sampling instants.

Problem 6   Compute the solution of the following differential equation:

$$\frac{d^2x(t)}{dt^2} + 4\frac{dx(t)}{dt} - 5x(t) = 0$$

in the interval $t \geq 0$, with initial conditions $x(0) = -1$ and $x'(0) = 2$.

Problem 7   Compute the solution of the following differential equation:

$$\frac{d^2x(t)}{dt^2} + 4\frac{dx(t)}{dt} + 3x(t) = 0$$

in the interval $t \geq 0$, with initial conditions $x(0) = 3$ and $x'(0) = -2$.

Problem 8   Compute the solution of the following differential equation:

$$\frac{d^2x(t)}{dt^2} + 6\frac{dx(t)}{dt} + 5x(t) = u(t)$$

in the interval $t \geq 0$, where $u(t)$ is the unit step function, and with initial conditions $x(0) = 0$ and $x'(0) = 0$.

Problem 9   Write the input-output transfer function for each of these systems described by the following differential equations ($u(t)$ is the input and $x(t)$ is the output):

$$\frac{dx(t)}{dt} + 4x(t) = u(t)$$

$$\frac{d^2x(t)}{dt^2} + 6\frac{dx(t)}{dt} + 5x(t) = u(t)$$

$$\frac{d^2x(t)}{dt^2} + 5\frac{dx(t)}{dt} + 8x(t) = \frac{du(t)}{dt} + 3\,u(t)$$

4 DIGITAL COMPUTER SIMULATION

4.1   Dynamic System Simulation

In Chapter 2 we saw how dynamic systems can be modeled by differential equations; this was followed in Chapter 3, with the development of solution methods. First, we suggested how this can be done analytically, and for a certain class of equations we gave formulas for the solution. Analytical solutions are exact. We then pointed out that numerical techniques can be used to compute approximate solutions. We presented one such technique, Euler's method, and used it to compute solutions to simple linear differential equations. We mentioned that these techniques can be applied to arbitrary linear differential equations. In fact, for many nonlinear differential equations for which it is difficult or impossible to provide analytical solutions, numerical techniques can still be employed to compute approximate solutions. These numerical techniques are particularly well suited for execution on digital computers.

In the context of automatic control, a differential equation describes the dynamic behavior of a system. One must solve this differential equation in order to determine exactly how the system operates. In this book, when we use a digital computer to compute the solution of some differential equation, we say that we are *simulating* the system, or performing a *simulation*. Not only can the computer perform the necessary computations, it can also display the results in graphs. Several software packages that have been produced over the last two decades include computer programs that allow one to perform simulations. Over the years these simulation packages have become quite sophisticated, powerful and very "user-friendly." The usefulness and importance of these software packages is undeniable, because they greatly facilitate the analysis and design of control systems. They provide a tremendous tool in the hands of control engineers. However, a word of caution must be sounded. The availability of such packages and the ease with which one can use them should in no way detract from learning the underlying concepts. Mastery of the theoretical foundation is a prerequisite for its correct implementation.

4.2   MATLAB/SIMULINK

MATLAB/SIMULINK is one of the most successful software packages currently available, and is particularly suited for work in control. It is a powerful, comprehensive and user-friendly software package for performing mathematical computations. Equally as impressive are its plotting capabilities for displaying information. In addition to the core package, referred to as

MATLAB, there are additional packages called "toolboxes," in several application areas. Specifically, there is one on control system design, another on optimization, another on symbolic math and several others. In this book we will not spend time learning how to use MATLAB, but rather will devote time to learning SIMULINK, a software package within MATLAB used for digital computer simulation. We will see that it is particularly easy to learn and use. Our objective in this chapter is to help the reader gain a basic understanding of this software package by showing how to set up and solve a number of examples.

The first step in the simulation process is to enter the data into the computer and give appropriate instructions for carrying out the computations. SIMULINK allows us to do this very easily, by manipulating graphical objects, drawing diagrams, entering data and setting some simulation parameters. Interested readers are encouraged to further explore this very complete and versatile mathematical computation package. For our exposition we use a PC with Windows 97 on which MATLB version 5.3 and SIMULINK version 3 have been installed.

Before we provide details of this process, it would be helpful to think of simulation at a conceptual level with the aid of a diagram (see Figure 4.1). We already have seen how such diagrams provide visual representations of dynamic system operation. We think of a system as taking its input, operating on it and generating a corresponding output. The functions $u(t)$ and $x(t)$ correspond to physical quantities. The $u(t)$ could be a force, an input voltage or inflow concentration, and $x(t)$ could be a position, an output voltage or an outflow concentration.

Figure 4.1
A Conceptual Block Diagram of the System

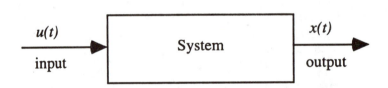

The simulation process at a conceptual level can also be thought of in "input-output" terms. The computer is provided with appropriate input data and other information about system structure, operates on this input data and generates output data, which it subsequently displays. A very nice feature of SIMULINK is that it exploits this interpretation and proceeds to visually represent the simulation process by using *simulink block diagrams*. Specifically, functions are represented by "subsystem blocks" that are then interconnected to form *simulink block diagrams* that define the simulation structure. Once the structure is defined, parameters are entered in the individual subsystem blocks that correspond to the given system data. Some additional simulation parameters must also be set to govern how the numerical computation will be carried out and how the output data will be displayed. We now describe the process in detail by numerically solving several differential equations using SIMULINK.

4.3   Examples Using SIMULINK

EXAMPLE
4.1

In Chapter 3 we computed the solution of the following differential equation, both analytically and numerically:

$$\frac{dx(t)}{dt} + x(t) = u(t) \qquad (4.1)$$

The input $u(t)$ was the unit step and the initial condition was $x(0) = 0$. Recall that the transfer function for this system is $T(s) = 1/(s + 1)$. We begin this section by using SIMULINK to numerically compute and display the solution of this differential equation in the interval $0 \le t \le 10$. We will proceed in two steps. We will first provide some general information about MATLAB/ SIMULINK and we will then construct a simulink block diagram for this differential equation.

To open MATLAB on a PC, first double-click the mouse button on the "shortcut" icon for MATLAB. If that is not available, click on the "start" location then highlight the "Programs" menu entry followed be the "MATLAB for Windows" menu entry and then highlight and click the "MATLAB" entry. A window opens up labeled "**MATLAB Command**" and the symbol "»" appears on a new line. Once in MATLAB, the next command to type is simulink (see Figure 4.2).

Figure 4.2
Command to
Invoke SIMULINK

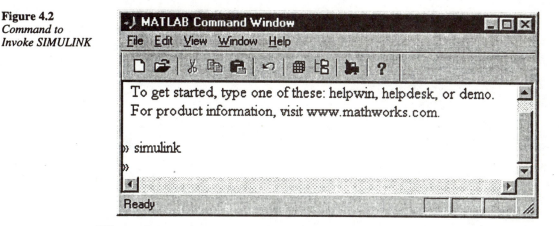

When the carriage return key is hit, a new window opens up labeled "**Simulink Library Browser**" (see Figure 4.3) which contains libraries of several subsystem blocks. If you click on the "+" sign next to the word "Simulink" a number of sub-libraries are revealed. These are labeled Continuous, Discrete, Functions & Tables, Math, Nonlinear, Signals &

Systems, Sinks and Sources. Each of these sub-libraries contains subsystem blocks that are used to construct simulink block diagrams. If you click on the "blank page" on the Simulink Library Browser menu a new window opens up labeled "**untitled**." It will be used to build up the particular simulink block diagram for our simulation.

Figure 4.3
Simulink Library Browser

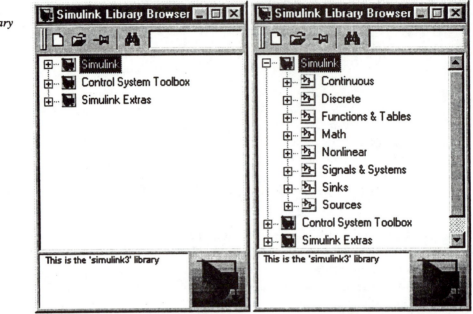

Clicking on the "+" sign next to "Souces" a number of subsystem blocks are revealed shown in Figure 4.4. Each of these subsystem blocks can be used to generate the corresponding function (action). For example, the subsystem block labeled "Step" is used to generate a step input, the subsystem block labeled "Sine Wave" can generate a sine function, and so forth. If one clicks on the Step block a description of this function appears at the bottom of the Simulink Library Browser.

Figure 4.4
Library of Sources

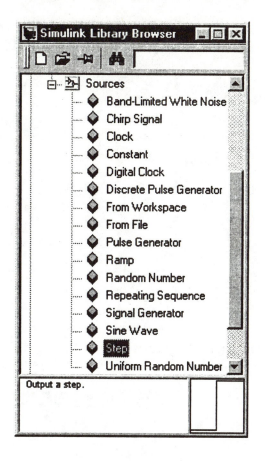

Clicking on the "+" sign next to "Sinks" in the Simulink Library Browser reveals the subsystem blocks shown in Figure 4.5. Each of these subsystem blocks can be used to perform the corresponding function. For example, the block labeled "Scope," can be used to generate a graph of a variable with the *x*-coordinate being "time." The subsystem block labeled "Display," displays the numerical value of some variable as time passes, and so forth. Since the subsystem block "Scope" is highlighted its description is displayed at the bottom of the Simulink Library Browser.

Figure 4.5
Library of Sinks

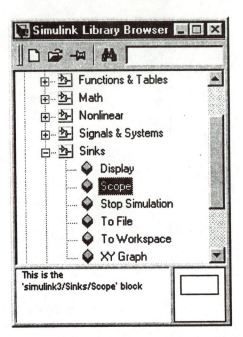

Clicking on the "+" sign next to "Continuous" in the Simulink Library Browser reveals a number of subsystem blocks shown in Figure 4.6.

Figure 4.6
Library of Linear Blocks

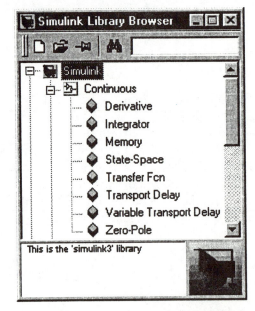

Each of these subsystem blocks can be used to specify dynamic characteristics of a particular system. For example, the subsystem block labeled "Transfer Fcn" can be used to specify the dynamic system model in terms of a transfer function.

These are some of the available subsystem libraries that contain basic building blocks. The reader is encouraged to explore the other libraries as well. These blocks are used to construct the appropriate simulink block diagram for a specific simulation. We are now ready to proceed to the next step, which is the construction of such a diagram for equation (4.1). To do this we use the **"untitled"** window (can be opened by clicking on the empty page entry in the Simulink Library Browser menu) shown in Figure 4.7. It will be used to build up an interconnection of simulink blocks from the subsystem libraries. The end result of this process will be the construction of a simulink block diagram for our differential equation.

Figure 4.7
A New Untitled
Window

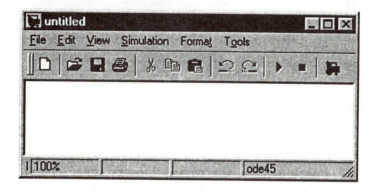

With the "Sources" sub-library open, move the pointer (cursor) and click on the block labeled "Step" and, while keeping the mouse button pressed down, drag the cube and place it inside the **"untitled"** window and release the mouse button. The result should look like Figure 4.8. Note that the icon has an arrowhead ">" pointing outward to designate an "output" interconnection port. This indicates that this block can only be connected to other blocks from this port.

Figure 4.8
Constructing a
Simulink Block
Diagram

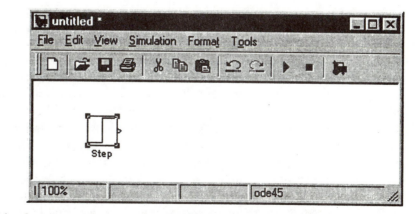

With the "Continuous" sub-library open, click on the block labeled "Transfer Fcn" and, while keeping the mouse button pressed down, drag it into the **"untitled"** window and release the mouse button (see Figure 4.9). The transfer function block has both an arrowhead pointing inward on one side and an arrowhead pointing outward on the other side. They designate input and output interconnection ports respectively. This means that if some input is placed in its input interconnection port the signal that will appear at its output interconnection port will be the output of a system with that transfer function when excited by the specified input (with zero initial conditions).

Figure 4.9
*Constructing a
Simulink Block
Diagram*

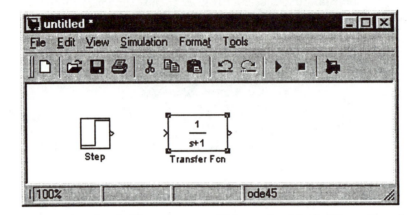

With the "Sinks" sub-library open, click on the block "Scope" and drag it into the "**untitled**" window (see Figure 4.10). The Scope block has an arrowhead pointing inward which designates that this is an input port.

Figure 4.10
*Constructing a
Simulink Block
Diagram*

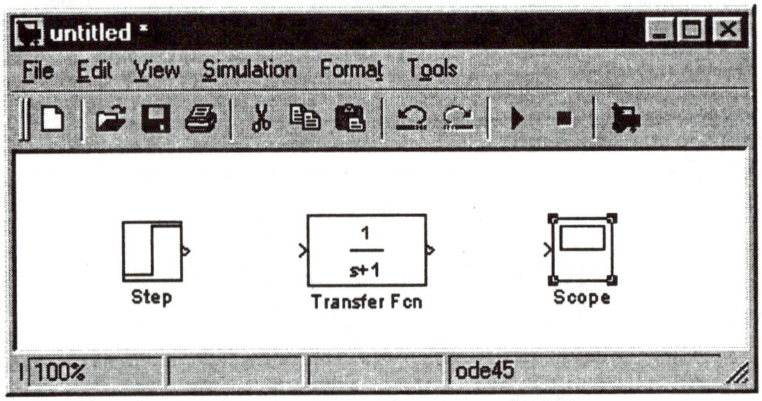

We have now completed the process of taking subsystem blocks from appropriate libraries and placing them in the "**untitled**" window. The next step is to interconnect these subsystem blocks and obtain the simulink block diagram. To do this we just need to work in the window labeled "**untitled**" and connect the three blocks. This is done by positioning the pointer on the output port of each block (pointer becomes "cross hairs") and dragging it to the input port of the next block. Specifically, we connect the output port of the "Step" block to the input port of the "Transfer Fcn" block. When the mouse button is released a solid arrow will appear connecting the two blocks (see Figure 4.11).

Figure 4.11
Constructing a
Simulink Block
Diagram

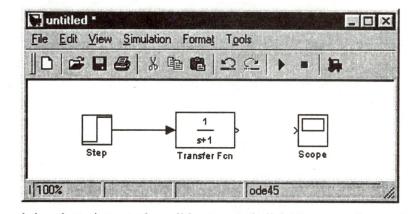

If we bring the pointer to the solid arrow and click the mouse button the solid arrow is shown in the "selected" state (see Figure 4.12). In this state we can remove the arrow by going to the "Edit" menu and dragging the pointer to the item "Cut" and releasing it (or by clicking on the "scissors") . Any object in the untitled window can be selected by placing the pointer on it and clicking the mouse button. It can then be cut in the same manner. Continuing the process, we bring the pointer to the output port (symbol ">") of the "Transfer Fcn" block and drag the arrowhead (now cross hairs) and position it at the input port (symbol ">") of the "Scope" block. When the mouse button is released, a solid arrow should appear connecting the two blocks (see Figure 4.13).

Figure 4.12
Constructing a
Simulink Block
Diagram

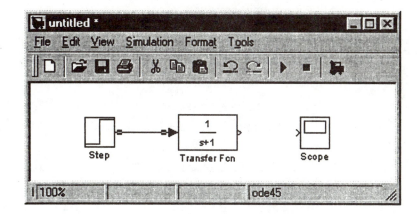

Figure 4.13
*A Simulink Block
Diagram*

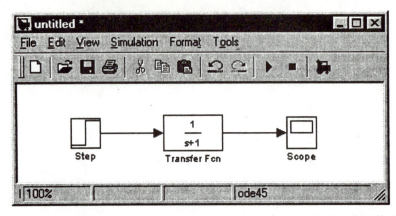

The three blocks have now been connected as desired and the simulink block diagram is complete. It shows that a step will be the input to a system described by some transfer function, and that the output will be computed and displayed in a graph. This is a generic diagram because we have not yet specified the particular system transfer function nor set additional parameter values. Notice the similarity between the simulink block diagram in Figure 4.13 and the conceptual block diagram in Figure 4.1.

Our next priority is to go into each of these blocks and set the parameters that correspond to our specific system. In addition, we need to set some simulation parameters. We begin with the step input by double-clicking on the block labeled "Step" in the "**untitled**" window. The dialog box in Figure 4.14 (it looks like a window) pops up.

Figure 4.14
*The Step Function
Dialog Box*

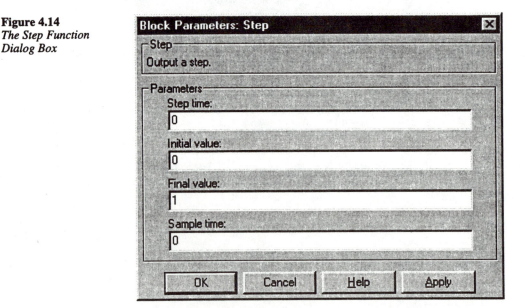

There are three parameters that need to be set: the step time (this is the time the value of the step changes from its initial value to its final value), the initial value and the final value of the step. The fourth parameter, Sample time, is the sample rate of the step which is "inherited" from the driven block. This gives us the flexibility to change these parameters. In our case, the step time needs to be set to zero (its default value is 1). This can be done by moving the pointer to that rectangle and placing it to the right of the digit to be changed and pressing the delete key. This erases the default value and allows us to enter the value 0 by typing it in (the changed Step time value is shown in Figure 4.14). The other two parameters need not be changed from their default values. When we are done, we click "OK."

Next we set the "Transfer Fcn" block parameters. Moving the pointer inside this block and double-clicking generates the dialog box shown in Figure 4.15.

In this window we enter the transfer function of our system. The numerator polynomial is defined by its coefficients. We enter these coefficients, each separated by a single space, starting on the left with the coefficient of the highest power. For example, the polynomial $s^2 + 3s + 2$ will be entered as [1 3 2] (left and right brackets are also needed). To change the default value (it is equal to 1) we move the pointer in the numerator rectangle and click the mouse button. We erase the default values (using the erase key) and type the required ones. We repeat the process with the denominator polynomial. *Our example has the default values, so in this case we do not need to change anything.* When we are done we click "OK."

There are no parameters to set for the "Scope." If we double-click the mouse button on the "Scope" block the window shown in Figure 4.16 pops up.

Figure 4.16
The Scope Dialog Box

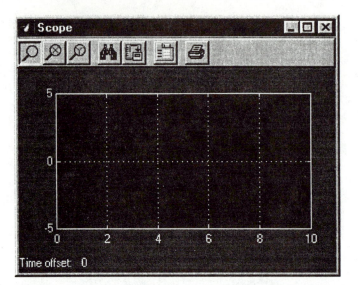

Finally, we need to set the parameters for the simulation run. We move the pointer to the menu labeled "simulation" and select "parameters." When we release the mouse button, a new dialog box pops up (see Figure 4.17).

Figure 4.17
The Simulation Parameters Dialog Box

There are a number of parameters that need to be set on the "Solver" card. At the top we specify the simulation "start time" and "stop time," which are set to 0 and 10 respectively. Next we set the simulation "Solver options." We choose a "Fixed-step" solver with a "step size" set equal to .1. The specific integration routine is the Euler method discussed in Chapter 3 which is denoted as "ode1 (Euler)." Figure 4.17 shows a completed simulation parameter dialog box. When we are done, we click "OK."

At this point in the process we have generated the appropriate simulink block diagram and entered the specific parameters for our system and simulation. We are now ready to execute the program and have the computer perform the simulation. First double-click on the "Scope" block so that it opens up. Then move the pointer to the "Simulation" menu of the "**untitled**" window and choose "Start." The Scope window shows the graph of $x(t)$, plotted on the y-axis, as a function of time. If we bring the pointer to the "binocular" icon (auto-scale toolbar button) of the Scope window and click, the graph will automatically be enlarged and displayed (see Figure 4.18). If at this point we want to obtain a hardcopy plot from the printer connected to the PC, we bring the pointer to the printer icon on the Scope window toolbar and click.

Figure 4.18
Step Response for
Equation (4.1)

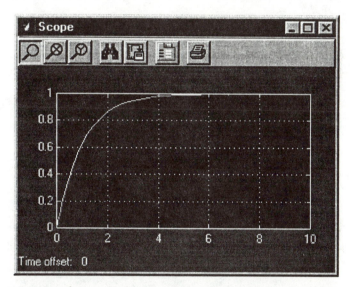

There is agreement between this graph and those obtained in Chapter 3 for the same example, where the exact solution was compared with the numerical solution (Figure 3.10) using the Euler method. In Chapter 3 we distinguished between approximate and exact solutions by indicating the approximate ones as $x_a(t)$. Here and in future chapters we will not make this distinction explicitly, as it will be understood from the context.

We used the above example to show how to enter data and carry out a simulation in the SIMULINK environment. The reader must agree that this is a very simple process. By following the same steps, we can use the procedure to numerically compute the solution of an arbitrary linear differential equation.

EXAMPLE
4.2

Consider now the following differential equation:

$$\frac{d^2x(t)}{dt^2} + 3\frac{dx(t)}{dt} + 2x(t) = 4u(t) \qquad (4.2)$$

where $u(t)$ is the unit step input and the initial conditions are $x'(0) = 0$ and $x(0)$ = 0. Again we would like to numerically compute the solution to this differential equation, in the interval $0 \le t \le 10$. The transfer function of this system is:

$$T(s) = \frac{4}{s^2 + 3s + 2}$$

Proceeding in exactly the same manner as above, we enter the data and generate the following simulink block diagram. In fact, we could have used the diagram generated in Example 4.1 and just changed the data in the subsystem blocks to reflect the new system (see Figure 4.19).

Figure 4.19
A Simulink Block Diagram for Equation (4.2)

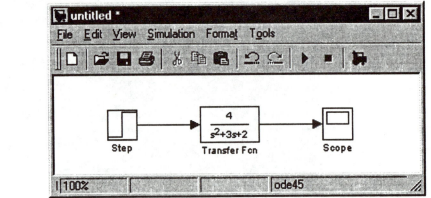

The only difference from the previous example is that the "Transfer Fcn" block now contains the new transfer function. The simulation parameters are left at the same values. The graph of the solution $x(t)$ is seen in Figure 4.20.

Figure 4.20
Step Response for
Equation (4.2)

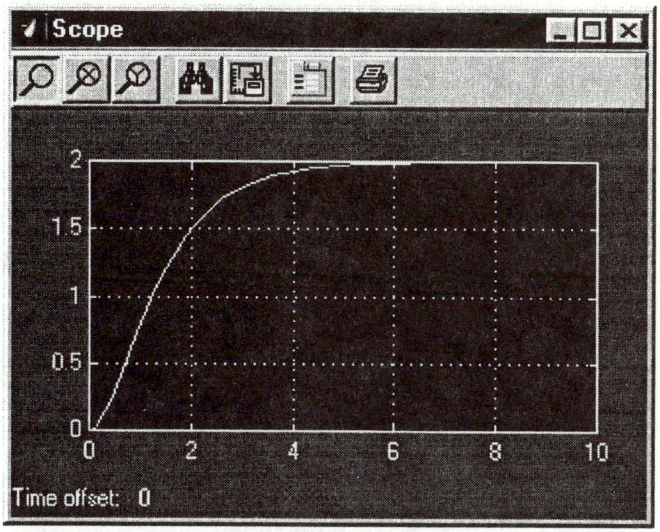

EXAMPLE 4.3

In both examples thus far, the initial conditions were set to zero. The next example shows how to construct a simulink block diagram when initial conditions are not equal to zero. Consider the differential equation (see equation (3.10)):

$$\frac{dx(t)}{dt} + x(t) = 0 \qquad (4.3)$$

with the initial condition $x(0) = 1$ and where we want to compute the solution over the interval $0 \leq t \leq 10$. We generate the simulink block diagram as before, with two modifications: (1) we use a different "system" block and (2) we have no "input" block. To proceed, we first need to write the differential equation in the so-called *state space* form. The general form looks like the following:

$$\frac{dx(t)}{dt} = Ax(t) + Bu(t) \qquad y(t) = Cx(t) + Du(t)$$

It consists of two equations, where $x(t)$ is the "state," $u(t)$ the "input" and $y(t)$ the "output." The first is called the *state* equation and the second the *output* equation. For equation (4.3), if we choose $y(t) = x(t)$ we can write:

$$\frac{dx(t)}{dt} = -1x(t) \quad y(t) = x(t)$$

Therefore, $A = -1$, $C = 1$ and since $u(t) = 0$ we let $B = 0$ and $D = 0$. In SIMULINK, rather than using a "Transfer Fcn" block to describe our system, we must use the "State-Space" block shown in Figure 4.21. Notice also that the "input" block is missing since $u(t) = 0$.

Figure 4.21
*A Simulink Block
Diagram for
Equation (4.3)*

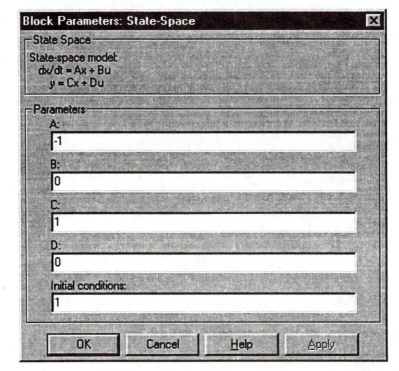

By double-clicking on the "State-Space" subsystem block the dialog box shown in Figure 4.22 opens up. The appropriate values are entered in the five parameter boxes labeled "A," "B," "C," "D" and "Initial Conditions" (i.e., $A = -1$, $B = 0$, $C = 1$, $D = 0$ and the initial condition = 1).

Figure 4.22
*The State-Space
Dialog Box for
Equation (4.3)*

Running the simulation by choosing the Euler method and setting the stop time to 10 with the step size set to .1 yields the graph shown in Figure 4.23. Compare this plot with that in Figure 3.1. Note that smaller step size values yield more accurate simulations (i.e., numerical differential equation solutions).

Choosing a "large" step size could lead to erroneous results!

Figure 4.23
The Solution of
Equation (4.3)

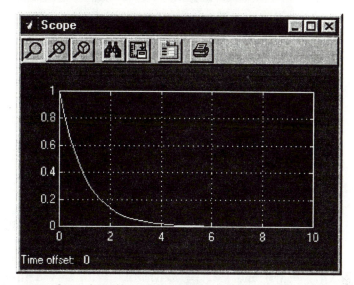

The state space form introduced in this example can be used to express and solve more general differential equations. The functions $x(t)$, $dx(t)/dt$, $u(t)$ and $y(t)$ are in general vectors and the variables A, B, C and D are appropriately sized matrices. The vector $x(t)$ is referred to as the "state" vector. We will see how this is done in the next example, where we first express the differential equation in state space form and then solve it. More details about this process can be found in the books listed in the bibliography.

EXAMPLE 4.4

Let us show how to set up a simulink block diagram to solve a second order differential equation with nonzero initial conditions and a step input. Consider the following differential equation:

$$\frac{d^2x(t)}{dt^2} + 3\frac{dx(t)}{dt} + 2x(t) = 4u(t) \qquad (4.4)$$

where $u(t)$ is the unit step function, the time interval of interest is $0 \le t \le 15$, and the initial conditions are $x(0) = 1$ and $x'(0) = -2$. We will again have to use a "State-Space" block in order to define our system because the initial conditions are not zero. The simulink block diagram for this system is given in Figure 4.24.

Figure 4.24
*A Simulink Block
Diagram for
Equation (4.4)*

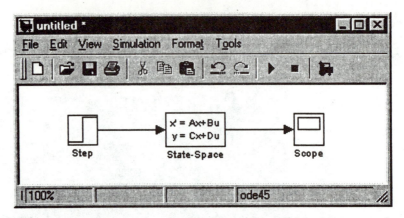

In order to enter the parameters in the "State-Space" block, we need to first express the differential equation in state space form. This can be done in many ways and the interested reader is directed to the bibliography for a more complete exposition. The "state" vector in this example will have two components, $x_1(t)$ and $x_2(t)$, each giving rise to a state equation. We define $x_1(t) = x(t)$ and $x_2(t) = dx_1(t)/dt$. With this definition, the given initial conditions can now be expressed as $x_1(0) = 1$ and $x_2(0) = -2$. The first state equation is then

$$\frac{dx_1(t)}{dt} = x_2(t)$$

The second state equation comes from substituting $x_1(t)$ and $x_2(t)$ in equation (4.4). Since $x_2(t) = dx_1(t)/dt$ and $dx_2(t)/dt) = d^2x(t)/dt^2$, we can rewrite equation (4.4) as:

$$\frac{dx_2(t)}{dt} + 3x_2(t) + 2x_1(t) = 4u(t)$$

The state space form of equation (4.4) then becomes:

$$\frac{dx_1(t)}{dt} = x_2(t)$$

$$\frac{dx_2(t)}{dt} = -2x_1(t) - 3x_2(t) + 4u(t)$$

In matrix form this is written as:

$$\begin{bmatrix} \dfrac{dx_1(t)}{dt} \\ \dfrac{dx_2(t)}{dt} \end{bmatrix} = \begin{bmatrix} 0 & 1 \\ -2 & -3 \end{bmatrix} \begin{bmatrix} x_1(t) \\ x_2(t) \end{bmatrix} + \begin{bmatrix} 0 \\ 4 \end{bmatrix} u(t), \quad y(t) = \begin{bmatrix} 1 & 0 \end{bmatrix} \begin{bmatrix} x_1(t) \\ x_2(t) \end{bmatrix}$$

where

$$A = \begin{bmatrix} 0 & 1 \\ -2 & -3 \end{bmatrix}, \quad B = \begin{bmatrix} 0 \\ 4 \end{bmatrix}, \quad C = \begin{bmatrix} 1 & 0 \end{bmatrix} \quad \text{and} \quad D = 0$$

For an arbitrary second order differential equation with a step input, these matrices will have the same structure. The first row of the A matrix will always be [0 1]. The second row will have entries that are the negative of the coefficients of the terms $x(t)$ and $dx(t)/dt$, respectively, in the differential equation. The B matrix will be a column vector with a zero first entry and the coefficient of $u(t)$ as the second entry. The C matrix will be [1 0], and $D = 0$. These are precisely the matrices needed as data for the simulink "State-Space" block, shown in the dialog box of Figure 4.25.

Figure 4.25
The State-Space Dialog Box for Equation (4.4)

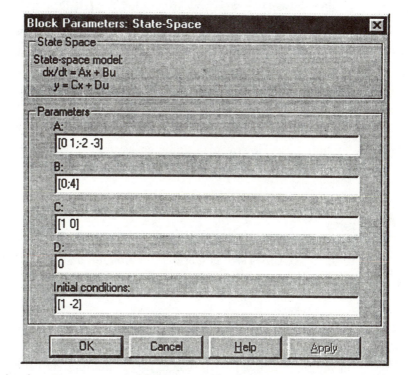

Matrix elements are entered by rows, with each row separated by a semicolon. Entries in each row are separated by a single space. The initial conditions are entered as shown in the corresponding block, with $x_1(0) = 1$ and $x_2(0) = -2$. The "Step" parameters must be changed to reflect that the step is at time $t = 0$. Setting the corresponding parameters for the simulation with step size equal to .1 yields the graph shown in Figure 4.26.

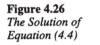
Figure 4.26
The Solution of
Equation (4.4)

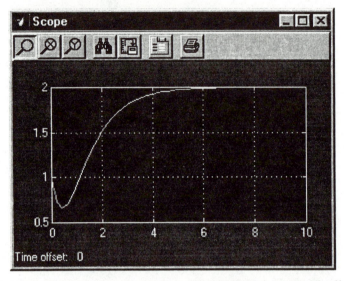

In this chapter we saw how to employ SIMULINK in numerically solving a number of first and second order differential equations. Even though we have just worked with simple differential equations, the techniques and software can be used to solve far more complicated examples. SIMULINK has many more features and capabilities than those introduced here and for a more complete description the reader is referred to the SIMULINK Version 3 documentation.

In all the examples we used the Euler method to numerically compute the solutions. We did so, because this method is the simplest to explain (see Section 3.2). However, in doing so we traded accuracy for simplicity. For a given step size there are other, more accurate methods including the Runge-Kutta (e.g., ode4).

4.4   Exercises

Problem 1   Use SIMULINK to numerically compute and graph the solution of the differential equation:

$$\frac{dx(t)}{dt} + 3x(t) = u(t)$$

in the interval $0 \le t \le 10$, where $u(t)$ is the unit step function and with initial condition $x(0) = 0$. Compute the exact solution and compare.

Problem 2   Use SIMULINK to numerically compute and graph the solution of the differential equation:

$$\frac{dx(t)}{dt} + 8x(t) = u(t)$$

in the interval $0 \leq t \leq 5$, where $u(t)$ is the unit step function and with initial condition $x(0) = 0$. Compute the exact solution and compare.

Problem 3   Use SIMULINK to numerically compute and graph the solution of the differential equation:

$$\frac{dx(t)}{dt} + 2x(t) = 0$$

in the interval $0 \leq t \leq 10$, with initial condition $x(0) = -1$. Compute the exact solution and compare.

Problem 4   Use SIMULINK to numerically compute and graph the solution of the differential equation:

$$\frac{dx(t)}{dt} - 1x(t) = 0$$

in the interval $0 \leq t \leq 10$, with initial condition $x(0) = 1$.

Problem 5   Use SIMULINK to numerically compute and graph the solution of the differential equation:

$$\frac{d^2x(t)}{dt^2} + 2\frac{dx(t)}{dt} + 10x(t) = 10u(t)$$

in the interval $0 \leq t \leq 8$, where $u(t)$ is the unit step function and initial conditions are zero. Compute the exact solution and compare.

Problem 6   Use SIMULINK to numerically compute and graph the solution of the differential equation:

$$\frac{d^2x(t)}{dt^2} + .1\frac{dx(t)}{dt} + 1x(t) = -u(t)$$

in the interval $0 \leq t \leq 60$, where $u(t)$ is the unit step function and initial conditions are zero. Compute the exact solution and compare.

5 STABILITY AND PERFORMANCE

5.1 Stability

In Chapter 2 we saw how physical laws provide the means for the development of dynamic models for systems. In many cases these models are given in terms of simple differential equations, which in Chapter 3 we solved both analytically and numerically. In Chapter 4 we computed numerical solutions of differential equations using SIMULINK. Our objective thus far was to characterize the solution exactly (i.e., in a *quantitative* manner). In this chapter we will investigate the dynamic behavior of systems and characterize it in a *qualitative* manner. This means that we do not want a characterization in terms of specific values, but rather a qualitative description. For example, we can give a quantitative description of some function, say $f(t) = t$ in the interval $0 \leq t \leq 10$, or merely give an imprecise qualitative description that the function is increasing in this range. We can say that $f(t) = \sin(t)$ (quantitative), or just say that the function oscillates (qualitative).

A very important qualitative characteristic of a dynamic system is *stability* (and instability). We all have a pretty good understanding of the concept of stability and are able to identify such behavior in physical systems. We have all at one time or other tried to balance a broomstick (inverted pendulum). After a few trials and failures, we usually can manage to balance it for a while by mastering the required back and forth moves. However, the broomstick (think of it as a system) cannot balance itself without our intervention. If we set it on the ground, balance it, and then let it go, it falls down. We would certainly identify this as unstable behavior. We have all seen on television early unsuccessful launches of rockets in the forties and fifties, where the rocket lifts off for a few feet and then tumbles and crashes. We would again associate this with unstable behavior. On the other hand, the dynamic behavior of a properly maintained water storage tank system in a modern bathroom would be characterized as stable, as the water does not overflow the tank. The same would be said of the dynamic behavior of a well-functioning boiler system in a large apartment building, in that its operating temperature remains within acceptable limits.

What we would like to do is give a mathematical definition that captures this physical interpretation of stability (and instability). Suppose that the operation of some dynamic system is described by some differential equation, as we saw in Chapter 2. If the system is stable, what should be the characteristics of the solution of this differential equation? If the system is unstable, what should characterize its solution? In particular, suppose that a system model is described in terms of a second order differential equation with constant coefficients. We know how to compute the solution and we know that the solution depends on the coefficients of the characteristic polynomial. The solution involves exponential terms and its shape is determined by these terms. It is interesting to examine what these terms look like and how they affect the solution. Suppose that the characteristic polynomial is equal to $s^2 + 3s + 2$, which has roots $r_1 = -1$ and $r_2 = -2$. Then the solution of the corresponding differential equation will contain the terms e^{-t} and e^{-2t}. One can easily see that the qualitative behavior (or shape) of both terms is that they are both "decaying" with time. In contrast, suppose that the characteristic polynomial was equal to $s^2 - 3s + 2$, which has roots $r_1 = +1$ and $r_2 = +2$. Then the solution of the corresponding differential equation will contain the terms e^t and e^{2t}. In this case both these terms "increase" with time, which means that the solution "blows up." Similar arguments can be made for systems described by first order differential equations with constant coefficients. A characteristic polynomial equal to $s + 3$, which has a root at -3, will correspond to a solution of the form e^{-3t}, which decays with time. On the other hand, a characteristic polynomial equal to $s - 4$, which has a root at 4, will correspond to a solution of the form e^{4t}, which increases with time.

In the case of differential equations with characteristic polynomials that have complex roots the same situation arises. Specifically, suppose that the characteristic polynomial is equal to $s^2 + 1s + 2.5$, which has roots $r_1 = -.5 + j1.5$ and $r_2 = -.5 - j1.5$. The solution will contain the terms $e^{-.5t}\cos(1.5t)$ and $e^{-.5t}\sin(1.5t)$. Again we see that the term that "dominates" is the exponential, where in this case it makes the terms "decay" with time, with some oscillation. Had the roots been $r_1 = +.5 + j1.5$ and $r_2 = +.5 - j1.5$, then the solution would contain the terms $e^{.5t}\cos(1.5t)$, and $e^{.5t}\sin(1.5t)$, which correspond to oscillations with "increasing" amplitude.

Clearly, there is a fundamental difference between the two situations. The solution of a differential equation describes the dynamic behavior of some physical variable $x(t)$. This variable represents some physical quantity like position, velocity, temperature, pressure, and the like, and it is important for us to know the qualitative characteristics of this quantity. Having a system variable that increases as a function of time without bound is certainly not an acceptable behavior and it is precisely this behavior that we would associate with system "instability." On the other hand, having system variables take values that are bounded and well behaved (nothing blows up) would be associated with system "stability."

It is important that we make a distinction between these two very different qualitative behaviors. This leads us to make the following "mathematical" definitions, which capture our physical notion of system stability and instability. Suppose a system is described by some differential equation with constant coefficients, or equivalently by the corresponding transfer function. We assume that the input (i.e., the right side of the equation) is either the zero function or some bounded function. For such a system stability, or instability, is determined by the location of the roots of the characteristic polynomial:

- If the characteristic polynomial has roots in the left half complex plane (i.e., their real part is negative) then we shall call the system *stable*. In this eventuality we also refer to the corresponding characteristic polynomial as being stable or having stable roots. The corresponding transfer function, which has this polynomial as its denominator, is also called stable.

- If the characteristic polynomial has roots in the right half complex plane, including the imaginary axis (i.e., their real part is zero or positive) we will call the system *unstable*. Similarly, we also refer to such a characteristic polynomial as being unstable or as having unstable roots. The corresponding transfer function, which has this polynomial as its denominator, is also called unstable.

EXAMPLE 5.1

Consider the single link robotic manipulator depicted in Figure 2.6, a system described by the following differential equation:

$$\frac{d^2x(t)}{dt^2} = u(t) \qquad (5.1)$$

with initial conditions $x(0) = 0$, $x'(0) = 0$ and where $u(t)$ is the unit step. Its characteristic polynomial is s^2, which means the system is *unstable*. We should expect things to "blow up." The solution, $x(t) = (1/2)t^2$, is plotted in Figure 5.1 and we immediately see that its value at time $t = 50$ is 1,250 and it continues to grow without bound.

Figure 5.1
The Solution of Equation (5.1)

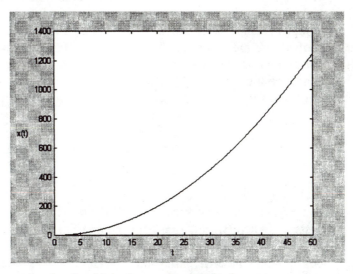

In this example, $J = 1$, $x(t)$ is the angle and the input torque is a step function. The initial conditions imply that the link starts at the zero angle position with zero angular velocity. However, its dynamic behavior is unstable. Its angular position and velocity increase without bound, even though the input to the system is bounded. The reason for this instability is not the fact that there is no "friction" in our system. The behavior would have been similar even if some frictional effects had been included in the model. In a physical robotic manipulator, the link is usually not allowed to rotate 360°, but is constrained by mechanical "stops." In that case we would see the link "banging into" these stops, which is clearly not acceptable either.

In this book we mainly deal with systems described by first or second order linear differential equations with constant coefficients. The corresponding characteristic polynomials will therefore be of degree one or two. However, the type of behavior we just discussed (stable or unstable) is also found in differential equations of higher order, with corresponding characteristic polynomials of higher degree. The definition of stability and instability given above applies to this class of systems as well. For systems described by dynamic models that are more general (e.g., nonlinear differential equations) similar definitions of stability and instability can be given that again capture our physical, intuitive understanding.

5.2   Performance

Often, in addition to stability, there are types of qualitative system behavior that we would like to characterize. Consider a system described by a second order differential equation with constant coefficients, which is stable. In other words, the roots of its characteristic polynomial are strictly in the left half complex plane. One can show that other attributes of dynamic system behavior will depend on the specific location of these roots inside the left half complex plane. Consider a second order differential equation, where all the coefficients are positive. This of course implies that the coefficients of the corresponding characteristic polynomial are positive. Specifically, let the characteristic polynomial be $s^2 + bs + c$, where $b > 0$ and $c > 0$. From the quadratic formula we can immediately determine that its roots are

$$r_1 = \frac{-b + \sqrt{b^2 - 4c}}{2} \quad , \qquad r_2 = \frac{-b - \sqrt{b^2 - 4c}}{2}$$

where regardless of the specific values of b and c, the roots are in the left half complex plane. Any such system will be stable. We consider this class of systems, where we also re-label the coefficients b and c by writing them as $2\zeta\omega_n$ and ω_n^2, respectively. In particular, let us consider the class of systems described by the differential equation and characteristic polynomial:

$$\frac{d^2x(t)}{dt^2} + 2\zeta\omega_n\frac{dx(t)}{dt} + \omega_n^2 x(t) = \omega_n^2 u(t), \qquad s^2 + 2\zeta\omega_n s + \omega_n^2$$

where ζ and ω_n are positive constants and $u(t)$ is the unit step input. This differential equation could be the model of several systems considered in Chapter 2 (e.g., the circuit in Figure 2.8). The parameter ζ is called the *damping ratio* and ω_n is called the *undamped natural frequency*. The reason we relabel in this way is because the ζ and ω_n so defined have special significance that will be apparent momentarily. The solution of this differential equation will be of the form $x(t) = x_h(t) + x_p(t)$ and $x_p(t) = 1$ (see Chapter 3). Since the system is stable, $x_h(t)$, which contains the exponential terms, will decay to zero. Therefore, after some time the solution will "tend" to 1. We can think of this as a *tracking problem*. We excite the system by a step input and want the output $x(t)$ to *track* the step input. In other words, we rate system performance by how well the output tracks the step input. From the quadratic formula the two roots are:

$$r_1 = \frac{-2\zeta\omega_n + \sqrt{4\zeta^2\omega_n^2 - 4\omega_n^2}}{2}, \quad r_2 = \frac{-2\zeta\omega_n - \sqrt{4\zeta^2\omega_n^2 - 4\omega_n^2}}{2}$$

If $0 = \zeta < 1$, then the two roots will be complex conjugate numbers (i.e., they have the same real part with the imaginary part being of the same magnitude but opposite sign). With $j = \sqrt{-1}$, let $\omega_d = \omega_n\sqrt{1-\zeta^2}$. Then $r_1 = -\zeta\omega_n + j\omega_d$ and $r_2 = -\zeta\omega_n - j\omega_d$. These roots are plotted in Figure 5.2.

Figure 5.2
Root Plot of Second Degree Polynomial

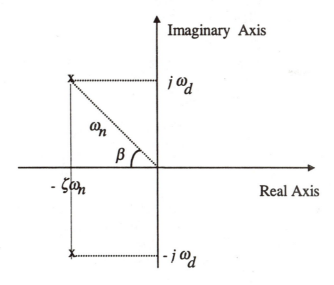

For the case when the value of ζ is in the range $0 \le \zeta < 1$, the solution and corresponding system are characterized as being *underdamped*. For the case in which the value of ζ is greater than or equal to 1, the roots are real numbers and the corresponding system and solution are called *overdamped*. We can make the following observations for the case of complex roots (i.e., $0 \leq \zeta < 1$) :

1) The value of ω_n is the distance of either root from the origin of the complex plane.

2) Suppose that ζ has a fixed value and we let ω_n vary. Then the angle β does not change. If we increase the value of ω_n, the position of either root will move farther away from the origin. If we decrease the value of ω_n, the roots move closer to the origin.

3) Suppose we fix the value of ω_n and let ζ vary. Then (a) the distance of the roots from the origin does not change (i.e., roots stay on a circular arc of radius ω_n); (b) if we increase the value of ζ the roots move closer to the real axis; (c) if we decrease the value of ζ the roots move closer to the imaginary axis.

It is interesting to examine the ramifications of characteristic polynomial root locations on the dynamic behavior, or performance, of second order systems. Rather than carrying out our discussion about a general system, it is instructive to fix our attention on specific examples by giving ζ and ω_n specific values. In all the examples that follow we will assume that the initial conditions are zero (i.e., $x'(0) = 0$ and $x(0) = 0$) and compute the unit step response $x(t)$ for $t \geq 0$. The "shape" of the step response would be the same if the step size is of some value other than 1.

EXAMPLE 5.2

Suppose first that the value of $\zeta = 2$ and $\omega_n = 1$. The corresponding differential equation is:

$$\frac{d^2x(t)}{dt^2} + 4\frac{dx(t)}{dt} + x(t) = u(t) \tag{5.2}$$

and the corresponding characteristic polynomial is:

$$s^2 + 4s + 1$$

with two real roots at -3.7321 and -.2679. This differential equation could be the model for the circuit example of Figure 2.8, with $L = 1$, $R = 4$, $C = 1$ and $u(t)$ the unit step function. The variable $x(t)$ would denote the voltage across the capacitor. From our work in Chapter 3 we know that the solution is made up of two parts: $x(t) = x_h(t) + x_p(t)$. In this case we have that $x_h(t) = fe^{-.2679t} + ge^{-3.7321t}$ and that $x_p(t) = a$, where a, f, g are constants. First we compute a by plugging this solution into the differential equation and making sure that the right side is equal to the left side for all t. Specifically, for any nonnegative t, we must have the left side a be equal to the right side 1. Therefore, $a = 1$.

Now with this information we use the initial conditions to determine the values of f and g. We must have that $x(0) = 0$, which implies that $f + g + 1 = 0$. Furthermore, we must have that $x'(0) = -.2679f - 3.7321g = 0$. Solving these two equations for f and g gives:

$$f = -1.0773, \quad g = .0773$$

Therefore, the solution of this differential equation for the time interval $t \geq 0$ is:

$$x(t) = -1.0773e^{-.2679t} + .0773e^{-3.7321t} + 1$$

It is interesting to plot this function and see how the solution evolves with time (see Figure 5.3 for a graph of this function in the range $0 \leq t \leq 30$). Notice, in particular, the shape of this curve. We know from the analytical expression of the solution that $x(t)$ starts at 0 at time $t = 0$ and as time increases the value tends to 1 (i.e., $\lim_{t \to \infty} x(t) = 1$). Clearly, it takes some 20 time units for $x(t)$ to get very close to 1, so it does not require a very long time for the solution to reach its "asymptotic" or "steady state" value. We can also see from the plot that $x(t)$ is increasing steadily, that there are no oscillations and that its value never exceeds unity. Since our performance objective is to have $x(t)$ track the unit step, we could say that it does reasonably well, but that it takes a somewhat long time for it to get to the level 1.

Figure 5.3
The Solution of Equation (5.2)

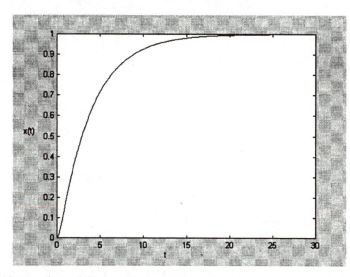

Note that if equation (5.2) was the model for the circuit example in Figure 2.8, the voltage of the capacitor would increase from 0 and settle down after some 20 units of time to the value 1.

EXAMPLE 5.3

Suppose now that the value of $\zeta = 2$ and $\omega_n = 4$. The corresponding differential equation is:

$$\frac{d^2x(t)}{dt^2} + 16\frac{dx(t)}{dt} + 16x(t) = 16u(t) \qquad (5.3)$$

and the corresponding characteristic polynomial is:

$$s^2 + 16s + 16$$

with two real roots at -14.928 and -1.0718. Again we point out that equation (5.3) could be the model for the circuit example in Figure 2.8, where $L = 1/8$, $C = 1/2$, $R = 2$ and the battery input is a unit step. The variable $x(t)$ represents the voltage across the capacitor. Proceeding as above we can compute the solution as:

$$x(t) = -1.0774e^{-1.0718t} + .0774e^{-14.928t} + 1$$

where a graph of this function in the range $0 \leq t \leq 30$ is shown in Figure 5.4.

Figure 5.4
The Solution of Equation (5.3)

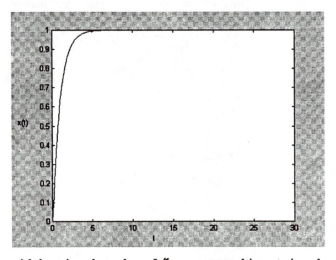

Notice that with keeping the value of ζ constant and increasing the value of ω_n we have made the response "faster," as it now reaches the value 1 in 6 time units. If equation (5.3) represented the circuit of Figure 2.8, with the different values of R, L and C, the capacitor voltage now reaches its "steady state" value much faster. Clearly, the tracking performance of this system is better than that of Example 5.3.

EXAMPLE
5.4

For our next example let us choose the values of $\zeta = 1.1$ and $\omega_n = 1$. The corresponding differential equation is:

$$\frac{d^2x(t)}{dt^2} + 2.2\frac{dx(t)}{dt} + x(t) = u(t) \qquad (5.4)$$

and the corresponding characteristic polynomial is:

$$s^2 + 2.2s + 1$$

with two real roots at -1.5583 and -.64174. Proceeding in exactly the same way as above we can compute the solution to be equal to:

$$x(t) = -1.7002e^{-.64174t} + .7002e^{-1.5583t} + 1$$

A plot of $x(t)$ for t in the range $0 \leq t \leq 30$ is shown in Figure 5.5.

Figure 5.5
*The Solution of
Equation (5.4)*

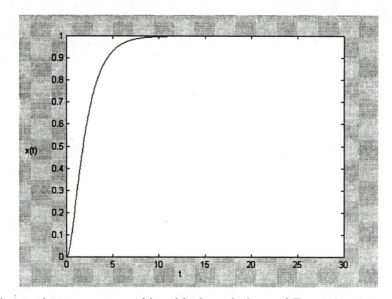

It is interesting to compare this with the solutions of Examples 5.2 and 5.3. The general shape of the solution is the same, but we immediately notice that the response is "faster" than that of Example 5.2 but not as fast as that of Example 5.3. It takes some 10 units of time to reach the value 1, compared with some 20 units of Example 5.2 and some 6 units of Example 5.3.

EXAMPLE
5.5

Suppose now that we consider a differential equation where the roots of the corresponding characteristic polynomial are complex numbers. Suppose that we let $\zeta = .7$ and $\omega_n = 1$. The corresponding differential equation is:

$$\frac{d^2x(t)}{dt^2} + 1.4\frac{dx(t)}{dt} + x(t) = u(t) \tag{5.5}$$

and the corresponding characteristic polynomial is:

$$s^2 + 1.4s + 1$$

with two complex conjugate roots at $-.7 + j.71414$ and $-.7 -j.71414$. From our discussion in Chapter 3, the solution of this differential equation has the form:

$$x(t) = ae^{-.7t}\cos(.71414t) + be^{-.7t}\sin(.71414t) + 1$$

where the constants a and b are computed from the initial conditions. The two equations in our case are $a + 1 = 0$ and $-.7a + .71414b = 0$. This implies that $a = -1$ and that $b = -.9802$. Therefore, the solution $x(t)$ is given below and plotted (see Figure 5.6) as a function of t in the range $0 \le t \le 30$:

$$x(t) = -e^{-.7t}\cos(.71414t) - .9802e^{-.7t}\sin(.71414t) + 1$$

Figure 5.6
The Solution of
Equation (5.5)

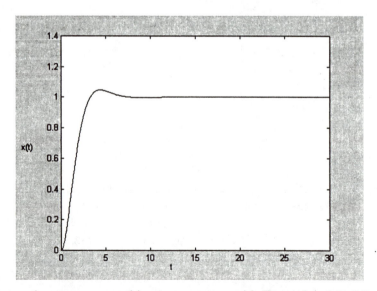

It is instructive to compare this step response with Examples 5.2, 5.3 and 5.4. The first difference to point out is that this solution has "overshoot." In other words, before the value settles to 1, for some period of time it overshoots the value 1. In fact, there is a slight oscillation about the value 1 (not very clear in this example). The other difference is that it settles to the value 1 faster than Examples 5.2 and 5.4, in 8 time units (compared with 10 and 20), but not as fast as Example 5.3 (6 units).

EXAMPLE
5.6

Finally, let us consider the case of $\zeta = .2$ and $\omega_n = 1$. The corresponding differential equation is:

$$\frac{d^2x(t)}{dt^2} + .4\frac{dx(t)}{dt} + x(t) = u(t) \qquad (5.6)$$

and the corresponding characteristic polynomial is:

$$s^2 + .4s + 1$$

with two complex conjugate roots at $-.2 + j.9798$ and $-.2 - j.9798$. Equation (5.6) could be the model for the circuit in Figure 2.8, with $R = .4$, $L = 1$ and $C = 1$. The solution has the form $x(t) = ae^{-.2t}\cos(.9798t) + be^{-.2t}\sin(.9798t) + 1$, where the constants a and b are computed from the initial conditions. The two equations in our case are $a + 1 = 0$ and $-.7a + .71414b = 0$, which implies that $a = -1$ and that $b = -.2041$. Therefore, the solution $x(t)$ is given below and plotted (see Figure 5.7) as a function of t in the range $0 \leq t \leq 30$.

$$x(t) = -e^{-.2t}\cos(.9798t) - .2041e^{-.2t}\sin(.9798t) + 1$$

Figure 5.7
The Solution of Equation (5.6)

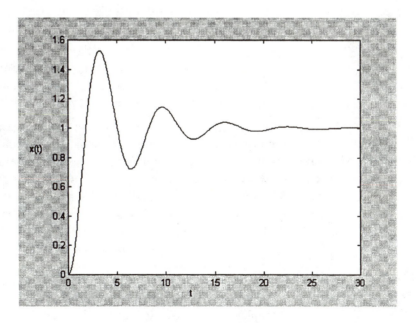

Clearly, the step response is quite oscillatory (i.e., it overshoots the value 1 then undershoots, and so on). Furthermore, the response is quite fast as it first reaches the value 1 in some two time units. However, the response does not stay at that value, but oscillates about it before it settles down. If equation (5.6) represented the circuit with the corresponding element values, the voltage of the capacitor would oscillate quite a bit. If equation (5.6) was the model of the spring-mass-dashpot system in Figure 2.4, then the solution would indicate that the mass would first oscillate before it settled down.

In the examples that were presented above we saw that the dynamic behavior of some stable dynamic system that is described by a second order differential equation can be characterized in terms of its response to the unit step function. We saw that its step response depended on the location of the roots of the characteristic polynomial. The step response is one of several performance measures that are used to characterize the dynamic behavior of a system. In view of the examples above, we can make the following observations:

- When $\zeta \geq 1$, the larger the value of ζ (for a fixed ω_n), the more "sluggish" the response.

- When $\zeta \geq 1$ and for a fixed value of ζ, when ω_n is increased the response becomes faster.

- When $0 \leq \zeta < 1$, the larger the value of ω_n (for a fixed ζ), the faster the response.

- When $0 \leq \zeta < 1$, the larger the value of ζ (for a fixed ω_n), the less oscillatory the response (i.e., the response is more "damped" and has less overshoot). However, a larger ζ implies slower response time.

Notice the conflicting performance outcomes in this last observation. For values of ζ in the range $0 \leq \zeta < 1$ and a fixed ω_n we can make the system respond faster by choosing ζ to be small, but at the same time the response would be more oscillatory (i.e., it would have more overshoot). This is one of the "classic" examples in control where a compromise must be reached.

We have just used mathematical expressions to characterize the behavior of stable dynamic systems when they are subjected to unit step inputs. It is important to have a clear understanding of what these expressions mean physically. Remember that solutions of differential equations describe how systems variables behave. The variable $x(t)$ could be a position, a velocity, a pressure or a temperature. The unit step input could be used to represent a level change in the input to a dynamic system. Specifically, the speed of a car is manually controlled at 40 mph but its cruise-control was previously set at 65 mph. At $t = 0$ the "resume" button is pushed. The thermostat setting in a house is set at some low temperature (furnace is on "stand-by") and at time $t = 0$ it is reset to 70^0 Fahrenheit. The output of such a system could then behave as in the examples given above (assuming that the system under consideration is described by such a differential equation). Consider now what it means for the output (step response) of our house heating system to be oscillatory as in Example 5.5. In the thermostat example this would mean that if the initial room temperature was 45° F, it would first shoot up to, say, 90° F, drop down to 55° F, go up to 80° F and come back to 65° F before it finally settled to 70° F. Clearly, this system performance would be unacceptable. Similar scenarios can be made for the dynamic behavior of the other types of systems (electrical, aerodynamic, chemical, financial, etc.).

In this chapter we dealt with only one aspect of system performance, namely, step response tracking. Many other performance specifications can be set on the dynamic behavior of systems. These may include the tracking of other types of signals (like ramp functions and sinusoids), bandwidth constraints (limitations on how fast the system can respond), the rejection of disturbance inputs (these are additional inputs to a system that cannot be controlled) or the attenuation of measurement noise (measurement errors introduced by the measurement process). Taking into account such performance specifications in control systems analysis and design leads to more effective automatic control. These issues are addressed in higher level control books (see Douglas, 1972, Ogunnaike and Ray, 1994, and Ogata, 1997).

5.3   Exercises

Problem 1   In each of the following cases, determine whether the system described by the corresponding differential equation model is stable or unstable:

$$\frac{dx(t)}{dt} + 5x(t) = 0$$

$$\frac{dx(t)}{dt} - x(t) = 0$$

$$-\frac{dx(t)}{dt} - x(t) = u(t)$$

$$\frac{d^2x(t)}{dt^2} + 2\frac{dx(t)}{dt} + 10x(t) = 10u(t)$$

$$\frac{d^2x(t)}{dt^2} - 7\frac{dx(t)}{dt} + 9x(t) = 10u(t)$$

Problem 2 In each of the following cases, determine whether the system described by the corresponding transfer function model is stable or unstable:

$$\frac{1}{s + 10}$$

$$\frac{-4}{s + 12.5}$$

$$\frac{2}{s - 3}$$

$$\frac{s + 2}{s^2 + 6s + 4}$$

$$\frac{s - 1}{3s^2 + 5s + 1}$$

$$\frac{3s + 1}{s^2 - 2s - 1}$$

Problem 3 Classify as overdamped or underdamped the solution (step response) of the system described by the second order differential equation model:

$$\frac{d^2x(t)}{dt^2} + 5\frac{dx(t)}{dt} + 1x(t) = 1u(t)$$

Use SIMULINK to obtain the step response when initial conditions are zero ($0 \leq t \leq 20$).

Problem 4 Classify as overdamped or underdamped the solution (step response) of the system described by the second order differential equation model:

$$\frac{d^2x(t)}{dt^2} + .3\frac{dx(t)}{dt} + 1x(t) = 1u(t)$$

Use SIMULINK to obtain the step response when initial conditions are zero ($0 \leq t \leq 30$).

6 FEEDBACK

6.1 Feedback Versus Open Loop Systems

From our discussion thus far we know that many dynamic systems can be modeled in terms of differential equations. We have learned how to obtain solutions to some basic differential equations that describe system behavior in a quantitative manner. Not only were we able to provide analytical expressions for the solution but we presented ways of computing (approximate) solutions numerically (e.g., the Euler method). SIMULINK was used to perform simulations. In Chapter 5 we saw how the dynamic behavior of system models can be characterized in a qualitative manner in terms of the notion of stability. For stable system models we examined system behavior even further in terms of step response tracking and made comments about the performance. In this chapter we would like to introduce the notion of feedback and see how its use can affect and dramatically improve system performance.

Our first priority is to gain a better understanding of what the word *feedback* means when it is used in the context of systems. At a very abstract level, systems are made up of components, each of which performs a function (action). These components are interconnected in some manner and this determines how the overall system operates. The operation of a system can be better understood if we represent it visually in terms of block diagrams, where each block represents a component and arrows that connect them specify the interconnections. At a conceptual level, it is also instructive to think of arrows as carrying information between components and system components as processing this information. We have repeatedly exploited this input-output characterization.

Consider how learning takes place in the school environment. What we present here is only one simple educational model on how learning can take place. The reader may not have thought in the past about this process in "system" terms, but it is helpful to do so. We focus on a system that is comprised of two components, a student and a teacher. The teacher receives completed homeworks, exams and projects (this is input to the "teacher" component). The teacher corrects the work and gives grades, provides verbal comments and completes report cards (this is the output of the "teacher" component). The student receives this information (i.e., this becomes the input to the "student" component) and then proceeds to complete new work, homework assignments, projects and exams (this is the output of the "student" component). In this process, the student uses the information provided by the teacher to adjust his/her study habits in order to improve "performance" in future assignments. We can visualize this operation by representing this learning process as in the following conceptual block diagram. The two system components are connected by arrows that indicate the flow of information (see Figure 6.1):

Figure 6.1
*Feedback
Interconnection*

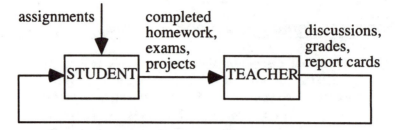

One can immediately recognize the presence of *feedback* in the operation of this system. It would also be instructive to examine a learning process in which no feedback was used. Specifically, we could construct a system model for learning with the same two components mentioned above (student and teacher) but that was structured as in Figure 6.2.

Figure 6.2
*Open Loop
Interconnection*

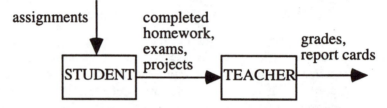

The teacher receives completed homework, exams and projects and corrects them and also assigns grades. These, however, are never returned to the student. The student just continues to complete and submit new work assignments to the teacher. Here no feedback is given to the student. The work is performed and grades and report cards are completed, but the student does not receive this information. In this case the system operates in an "open loop" fashion. One can certainly see the difference between the operation of these two learning systems and one can speculate that, when feedback is provided, then learning takes place more effectively.

Consider a different example of a health care system composed of two components, a patient and a doctor. The patient arrives at the doctor's office with some symptoms (this is output of the patient component). After an examination and some lab tests (this is input to the doctor component) the doctor makes an evaluation and provides feedback to the patient regarding his or her condition, frequently prescribing medication to correct the malady (this is output of the doctor component). The patient receives this information and medication (this is input to the patient component). If the diagnosis is accurate and the corrective actions successful, the patient recovers (improved performance). In some cases follow-up visits may be required with more exams and additional prescriptions. This again operates as a feedback system. One can also envision an open loop system operation in which the doctor examines the patient but says absolutely nothing to the patient and provides no medication. Another open loop operation would be to have the doctor prescribe a specific drug (say, aspirin) regardless of the results of the examination and send the patient home. Both these open loop operations are clearly less effective and may lead to catastrophic events.

Feedback is also used in the operation of a multitude of physical systems. This class of systems is very broad and encompasses biological, ecological, electrical, mechanical, chemical and financial systems. In some cases, the use of feedback is very evident and in others it may not be. In particular, humans use visual feedback in a variety of circumstances. In Chapter 1 we saw how feedback is used in the manual or automatic regulation of car speed. Visual feedback about the surroundings is also essential in the operation of a vehicle. Without this information, it is impossible for a driver to operate a vehicle safely. Consider the task of picking up a glass of water from a table and bringing it to your mouth. You first identify the location of the glass, move your hand toward it, pick it up–making sure not to tip it–and then bring it to your mouth. Information from the external world (spatial relationships in this case) is fed back to your brain (through your eyes), which makes it possible to successfully complete the task. If this information were not made available you would have great difficulty executing this task.

Feedback is also used in the operation of man-made systems. Consider the heating system in a typical house. It is composed of three components: the furnace, the thermostat and a comparator (frequently included in the thermostat housing). A desired temperature is set and the room temperature is measured. If the desired temperature is higher than the actual temperature, a signal is given to the furnace to turn on. The thermostat measures continually the room temperature and feeds back this information to the comparator. If the set temperature is reached, a signal is sent to the furnace to switch off. As heat is lost the temperature falls. This information is again fed back to the comparator, which now sends a signal for the furnace to be turned on. This process continues indefinitely. The diagram in Figure 6.3 depicts the operation of this system.

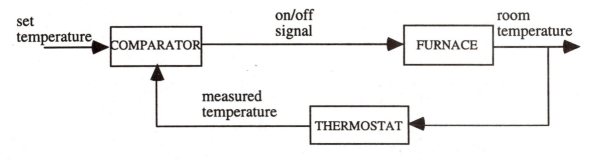

Figure 6.3 *Conceptual Block Diagram of House Heating System*

One can also envision the operation of a heating system in which no feedback is used. In particular, the measured temperature is not provided to a comparator. Rather, an open loop scheme is suggested (without using the room temperature measurement) for turning the furnace on and off. One such scheme could be: furnace comes on for twenty minutes and then is off for forty minutes and the pattern is repeated. The intervals could be adjusted for each month. It is not difficult to see that such an open loop operation would not regulate the room temperature very effectively. What would happen, for example, if a cold spell or a warm spell came? System performance would degrade.

6.2   Transfer Function Block Diagrams

Our discussion about feedback thus far was done at a conceptual level. Arrows in our feedback diagrams were used to depict the "flow of information" in a system and the blocks represented how system components manipulated this information. In this section we would like to become much more specific and quantitative. Our diagrams will actually depict specific mathematical relationships. Each block will correspond to a specific dynamic model of some system component, which relates its input to its output in a very particular manner. These inputs and outputs will be specific functions (which would represent physical variables). It turns out that if we employ transfer function models of systems, this results in a dramatic simplification. In particular, consider a system with a dynamic model given by some differential equation:

$$4\frac{d^2x(t)}{dt^2} + 2\frac{dx(t)}{dt} + 6x(t) = f(t) \qquad (6.1)$$

where $f(t)$ is thought of as the input and $x(t)$ is the output. As we have seen in Chapter 3, we can represent this system in an equivalent manner using its transfer function. Using the transformed variables $X(s)$ and $F(s)$ and the transfer function $G(s) = 1/(4s^2 + 2s + 6)$, we can write:

$$X(s) = \frac{1}{4s^2 + 2s + 6}\, F(s) \qquad (6.2)$$

This particular mathematical relationship can be expressed in the following *transfer function block diagram* in Figure 6.4.

Figure 6.4
Transfer Function
Block Diagram
Expressing
Equation (6.2)

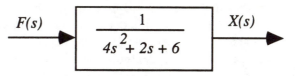

This block diagram not only says that there is a relationship between the output $X(s)$ and the input $F(s)$, it specifies this relationship completely in terms of expression (6.2). The output $X(s)$ is the product of $1/(4s^2 + 2s + 6)$ and $F(s)$. Suppose now that we have two system components, each of them described by its transfer function, and that they are connected in a particular manner. The mathematical relationships are:

$$X(s) = \frac{1}{s+1} F(s) \quad \text{and} \quad Y(s) = \frac{3}{s+2} X(s) \tag{6.3}$$

We have two system components, the first having a relationship between its input $F(s)$ and its output $X(s)$, given by the transfer function $G_1(s) = 1/(s + 1)$. The second has a relationship between its input $X(s)$ and its output $Y(s)$, given by the transfer function $G_2(s) = 1/(s + 2)$. Notice that the output of system component $G_1(s)$ is the input to system component $G_2(s)$. We can depict this "series" interconnection using the transfer function block diagram (see Figure 6.5).

Figure 6.5
A Series
Interconnection
Transfer
Function Block
Diagram
Expressing
Equation (6.3)

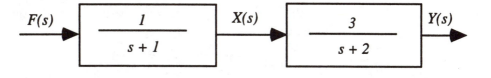

From the mathematical relationships in (6.3), the transfer function $T(s)$ of the overall system ($F(s)$ is the input, $Y(s)$ is the output) is given by:

$$Y(s) = \frac{3}{s+2} X(s) = \frac{3}{s+2} \frac{1}{s+1} F(s) = \frac{3}{s^2+3s+2} F(s) = T(s) F(s)$$

where $T(s) = 3/(s^2 + 3s + 2)$. If we are just interested in the overall system, we can depict it in the transfer function block diagram shown in Figure 6.6.

Figure 6.6
The Overall
Transfer Function
Block Diagram
for the Series
Interconnection

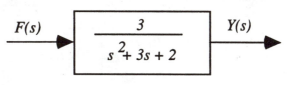

Recall that in Chapter 5 we defined stability of systems described by linear differential equations with constant coefficients or transfer functions. The overall series interconnection will be stable if each transfer function is stable. This makes the overall denominator a stable polynomial since it is the product of stable polynomials.

6.3 A Basic Feedback Interconnection

We have just seen how transfer function block diagrams can be used to depict particular mathematical relationships between system components that are interconnected in series. We would like to explore this further by incorporating feedback. Consider a system component that is described by a transfer function:

$$Y(s) = \frac{3}{s^2 + 3s + 2} E(s).$$

Suppose that the output $Y(s)$ of this system is measured and then subtracted from some other reference input $R(s)$. The resulting signal, $R(s) - Y(s)$, is then used as input to the same system (fed back):

$$E(s) = R(s) - Y(s) \qquad (6.4)$$

We can depict this feedback interconnection with the transfer function diagram shown in Figure 6.7.

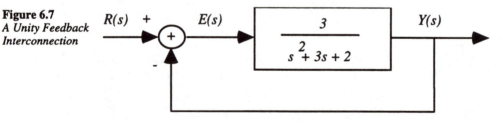

Figure 6.7
A Unity Feedback Interconnection

We would like to know the transfer function of the overall system. In other words, we want to know the overall transfer function from the input $R(s)$ to the output $Y(s)$. That is, find the $T(s)$ that relates $Y(s)$ and $R(s)$ by $Y(s) = T(s) R(s)$. After we compute this $T(s)$, we can depict the overall relationship as seen in Figure 6.8.

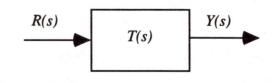

Figure 6.8
Overall Transfer Function of Unity Feedback Interconnection

We can obtain this by manipulating the mathematical relationships that are depicted in Figure 6.7. Namely, we substitute $E(s) = R(s) - Y(s)$ in the relationship $Y(s) = 3/(s^2 + 3s + 2) E(s)$. This gives:

$$Y(s) = \frac{3}{s^2 + 3s + 2} (R(s) - Y(s))$$

These are now algebraic expressions that we can manipulate using algebra. We can write:

$$Y(s) + \frac{3}{s^2 + 3s + 2} Y(s) = \frac{3}{s^2 + 3s + 2} R(s)$$

which implies

$$(1 + \frac{3}{s^2 + 3s + 2}) Y(s) = \frac{3}{s^2 + 3s + 2} R(s)$$

and

$$\frac{s^2 + 3s + 2 + 3}{s^2 + 3s + 2} Y(s) = \frac{3}{s^2 + 3s + 2} R(s)$$

And if we multiply both sides with $s^2 + 3s + 2$ and then divide by $s^2 + 3s + 5$ we have:

$$Y(s) = \frac{3}{s^2 + 3s + 5} R(s)$$

So the transfer function $T(s)$ of the overall system is given by:

$$T(s) = \frac{3}{s^2 + 3s + 5}$$

It should be clear that we could have performed these mathematical operations regardless of the specific form of the original system transfer function and developed the overall transfer function for a general system.

Let us now consider a slightly more involved feedback interconnection of system components (see Figure 6.9). The reader can immediately recognize that this interconnection structure is inspired by the discussion in Chapter 1. There we were interested in automating the process for controlling car speed.

We suggested a feedback interconnection inspired by how speed is controlled manually and introduced a specific control scheme. We will assume that the system under consideration, called the "plant," can be modeled by a transfer function and that we will use a controller that is also described by a transfer function. We will first give the mathematical expressions that describe the interconnection using transfer function models and then construct the corresponding transfer function block diagram that depicts it. Suppose that we have a system with input $F(s)$ and output $Y(s)$ and transfer function $P(s)$ (P stands for "plant"). This transfer function, which is a rational function in s (the ratio of two polynomials) has a numerator polynomial $n_p(s)$ and a denominator polynomial $d_p(s)$. For a specific plant, the numerator and the denominator will be polynomials with given coefficients. Here we would like to consider the general case and develop general formulas:

$$P(s) = \frac{n_p(s)}{d_p(s)}, \qquad Y(s) = P(s)\,F(s) \tag{6.5}$$

Suppose that we can measure the output signal and that we subtract it from some reference input $R(s)$:

$$E(s) = R(s) - Y(s) \tag{6.6}$$

We then let $E(s)$ be the input to some other system component with transfer function $C(s)$ (C stands for "controller"), whose output is then set to be the "plant" input. Specifically,

$$F(s) = C(s)E(s)$$

where

$$C(s) = \frac{n_c(s)}{d_c(s)} \tag{6.7}$$

Again we see how the output of the plant is measured and fed back to the input of the plant (via the controller). This interconnection is depicted in the transfer function block diagram of Figure 6.9.

Figure 6.9
The Standard
Unity Feedback
Interconnection

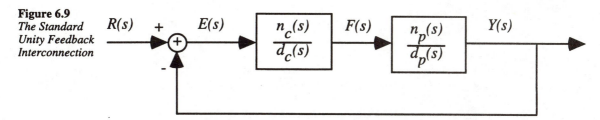

Suppose that we now want to develop the transfer function of the overall system, with input $R(s)$ and output $Y(s)$. Specifically, we want to write down the transfer function $T(s)$, which would relate $R(s)$ and $Y(s)$ $(Y(s) = T(s) R(s))$ and is shown in Figure 6.10.

Figure 6.10
The Overall
Transfer Function

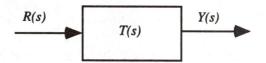

To do this we manipulate the mathematical expressions that define this interconnection and are shown in Figure 6.9.

$$Y(s) = \frac{n_p(s)}{d_p(s)} \, F(s), \quad F(s) = \frac{n_c(s)}{d_c(s)} \, E(s)$$

$$E(s) = R(s) - Y(s)$$

It is clear that:

$$Y(s) = \frac{n_p(s)}{d_p(s)} \, F(s) = \frac{n_p(s)}{d_p(s)} \frac{n_c(s)}{d_c(s)} \, E(s) = \frac{n_p(s)}{d_p(s)} \frac{n_c(s)}{d_c(s)} (R(s) - Y(s))$$

This implies that:

$$(1 + \frac{n_p(s)n_c(s)}{d_p(s)d_c(s)}) Y(s) = \frac{n_p(s)n_c(s)}{d_p(s)d_c(s)} \, R(s)$$

$$\frac{d_p(s)d_c(s) + n_p(s)n_c(s)}{d_p(s)d_c(s)} \, Y(s) = \frac{n_p(s)n_c(s)}{d_p(s)d_c(s)} R(s)$$

multiplying both sides by $d_p(s)d_c(s)$ and dividing by $d_p(s)d_c(s) + n_p(s)n_c(s)$, yields:

$$Y(s) = \frac{n_p(s)n_c(s)}{d_p(s)d_c(s) + n_p(s)n_c(s)} \; R(s)$$

So the overall transfer function for this feedback interconnection, also called the *closed loop transfer function*, is given by:

$$T(s) = \frac{n_p(s)n_c(s)}{d_p(s)d_c(s) + n_p(s)n_c(s)} \tag{6.8}$$

Note that the denominator of this transfer function is the polynomial:

$$\phi(s) = d_p(s)d_c(s) + n_p(s)n_c(s)) \tag{6.9}$$

This is called the *"closed loop" characteristic polynomial* of the overall system. Following our discussion in Chapter 5 about stability, this closed loop system will be stable if the closed loop characteristic polynomial is stable. Therefore, the overall system with transfer function $T(s)$ will be stable if the polynomial $\phi(s)$ in equation (6.9) is stable.

6.4   Use of Feedback for System Stabilization

In the previous section we developed general expressions for a basic feedback interconnection. If we are given a particular plant described by its transfer function and use that feedback interconnection with a specific controller, then the overall system would be characterized by a new transfer function. In other words, the dynamic behavior of the overall system would now be governed by a different transfer function and as a consequence a different differential equation. This opens up many possibilities. The first question is: To what extent can we change the dynamic behavior of systems using feedback? This is a fundamental question and addresses the very core of what automatic control is all about. In the remaining part of this chapter we will begin to understand the power of feedback by examining specific ways in which feedback can be used to alter the dynamic behavior of a system. In this section we begin with what is referred to as *system stabilization*.

Recall that a system model described by a differential equation or transfer function is called stable if its denominator, the characteristic polynomial, has all its roots in the left half complex plane. We have already pointed out that stability is a very important system property because the lack of stability is accompanied by a dynamic behavior where signals blow up (very undesirable). A temperature or pressure blowing up in a chemical process would almost certainly lead to catastrophic results. The rotational speed of some mechanical system or engine blowing up would have similar consequences. Dynamic behaviors such as these should not be allowed and steps must be taken to ensure they never occur. We have already seen how the use of feedback affects dynamic behavior in a number of situations (e.g., student-teacher learning system, picking up object example, etc.). It is natural to inquire about the effects of feedback in the context of system stability. Specifically, can we use feedback to make a system stable? The answer to this question is a resounding yes! We will see that the unstable dynamic behavior of a system can be made stable by the use of feedback. In other words, feedback can be used to *stabilize* the overall system. Let us begin with a specific example.

Suppose that we have a plant modeled by the following differential equation:

$$\frac{dy(t)}{dt} - y(t) = f(t)$$

We can also say that it is described by the transfer function:

$$T(s) = \frac{1}{s - 1}$$

The characteristic polynomial of the system is $s - 1$, which has a single root equal to 1, that lies in the *right half complex plane*. The system is clearly *unstable* and we can see this unstable behavior if we solve the differential equation for some initial condition and a given function $f(t)$. Suppose that the initial condition is $y(0) = 0$ and that $f(t)$ is a unit step function. The solution of the differential can be computed using the techniques of Chapter 3 and is shown to be:

$$y(t) = e^t - 1 \quad \text{for} \quad t = 0$$

This can easily be verified by substitution back into the differential equation. Had $y(t)$ corresponded to some physical quantity (temperature, pressure, position, velocity, etc.) this expression shows that the system output would have increased without bound. Such behavior cannot be tolerated.

Suppose that we now place this system in the basic feedback configuration, with a controller $C(s)$, as depicted in Figure 6.11.

Figure 6.11
Placing Feedback Around an Unstable System

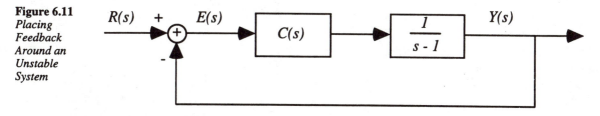

To physically implement this feedback scheme we need to be able to measure the system output $y(t)$ and have it "subtracted" from some other signal $r(t)$. This difference would then be used as the input to a new device, the controller $C(s)$, whose output then becomes the input to the plant. Furthermore, we need to choose, or more appropriately *design*, the controller (i.e., provide its transfer function). Clearly we have a great deal of freedom in this, but let us use the simplest possible choice. Let $C(s) = k$, a constant, which at this point should be thought of as a parameter whose value is to be chosen. In order to "plug into" the expressions developed in the previous section, we think of a constant as a ratio of polynomials, where the numerator is k and the denominator is equal to 1. With this controller the overall transfer function from expression (6.8) is:

$$T(s) = \frac{k}{s - 1 + k}$$

and the closed loop characteristic polynomial is equal to:

$$\phi(s) = s - 1 + k$$

The parameter k can be chosen freely. We can immediately notice that if we choose k to have a value greater than 1, then the closed loop characteristic polynomial will have its only root in the left half complex plane. In other words, the overall system will be stable. This is rather remarkable. Our given plant was unstable, with unacceptable dynamic behavior, but if the appropriate controller is chosen with feedback, the overall system can be made stable (i.e., the overall system can be *stabilized*). This is a direct consequence of the use of feedback, and demonstrates its power.

For this example we can see that a very simple controller, just a constant, was sufficient to stabilize the overall system. It is important to note that the differential equation of the plant was only first order. What would happen if the plant was described by a second order differential equation? In this case the plant transfer function would have a denominator polynomial of degree two. Would the same stabilization scheme work? Specifically, suppose that the given plant has the dynamic representation:

$$\frac{d^2y(t)}{dt^2} = f(t) \tag{6.10}$$

with the corresponding transfer function:

$$T(s) = \frac{1}{s^2} \tag{6.11}$$

This system could be the model of the single link robotic manipulator discussed in Chapter 2. In Chapter 5 we saw that this was an unstable system. We place it in the basic feedback configuration (see Figure 6.12).

Figure 6.12
Unity Feedback Configuration with Constant Controller

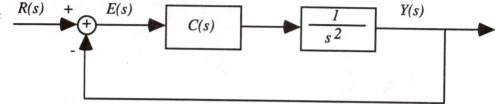

There we again choose $C(s) = k$, where k a real constant. The closed loop characteristic polynomial of the overall system from (6.9) will now be:

$$\phi(s) = s^2 + k$$

We need to stabilize the system. That is, we need to find, if possible, a value for k such that both roots of this characteristic polynomial are in the left half complex plane. From the quadratic formula we have that the roots are:

$$r_1 = \sqrt{-k}, \qquad r_2 = -\sqrt{-k}$$

Let us investigate what happens. The quantity under the radical is equal to $-k$. Suppose that $k > 0$. Then the roots are complex lying on the imaginary axis. For $k = 0$, they both lie at the origin. If $k < 0$, the roots are real, with one being positive and the other negative. Therefore, regardless of what value we give to k, the system will always have at least one unstable root. For this example we see that with a constant controller we cannot stabilize the system. This is not a "special case," as many such second order systems cannot be stabilized with a constant controller.

We can make some observations from this exercise. We have a single parameter k, and we want to affect the location of two roots. It would seem that we do not have enough "degrees of freedom" to perform this task. Suppose that we could introduce an additional parameter in the characteristic polynomial, in the coefficient of the term s. Consider the polynomial:

$$s^2 + k_1 s + k_2 \tag{6.12}$$

which we are asked to stabilize by choosing the parameters k_1 and k_2. The problem now becomes much easier. In fact, we can not only make this polynomial stable, we can make it have any roots we want. Suppose we want to make it have the roots -3 and -5. Expressed differently we want to make the polynomial in (6.8) equal to a polynomial that has these two roots. One such polynomial is:

$$(s + 3)(s + 5) = s^2 + 8s + 15 \qquad (6.13)$$

Clearly, if we choose $k_1 = 8$ and $k_2 = 15$, our task would be completed, as the polynomial in (6.12) becomes equal to (6.13). Observe that in this example we have two free parameters to affect the location of two roots, and we are able to do it. An immediate question arises: How can we introduce these "additional" parameters in the closed loop characteristic polynomial? The answer is to use more complex controllers that have transfer functions with more parameters.

The next level of complexity would be to use a controller with a transfer function where both its numerator and denominator polynomials are degree one polynomials. Suppose that we use a controller with transfer function:

$$C(s) = \frac{k_1 s + k_2}{s + k_3}$$

where k_1, k_2, and k_3 are "design" parameters. Having the ability to pick three parameters gives us more freedom in the design. Let us use such a controller for the system in equation (6.10). The system is described by:

$$\frac{d^2 y(t)}{dt^2} = f(t)$$

with corresponding transfer function:

$$T(s) = \frac{1}{s^2}$$

Let us place this system in the basic feedback configuration with controller:

$$C(s) = \frac{k_1 s + k_2}{s + k_3}$$

This implies that the overall feedback system will be depicted as in Figure 6.13.

Figure 6.13
Feedback
Configuration
with First
Order
Controller

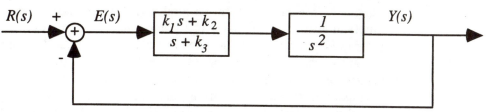

The closed loop characteristic polynomial of this system (from 6.9) is:

$$\phi(s) = (s + k_3)s^2 + (k_1 s + k_2) = s^3 + k_3 s^2 + k_1 s + k_2 \qquad (6.14)$$

The closed loop characteristic polynomial has degree three and will have three roots. We can readily find values for the three parameters that make the roots be -1, -2 and -3. One polynomial with these three roots is:

$$(s + 1)(s + 2)(s + 3) = s^3 + 6s^2 + 11s + 6 \qquad (6.15)$$

Therefore, we will complete our task if we find values k_1, k_2 and k_3 such that the polynomials in (6.14) and (6.15) are made to be the same. These values are:

$$k_3 = 6, \quad k_1 = 11, \quad k_2 = 6$$

If our system were described by a different second order differential equation, this procedure would lead to a more complicated set of three linear equations in three unknowns. In our example the solution was easily obtained.

In this example we were able to design a controller that stabilized the overall system. If the system in (6.10) was the model for a single link robotic manipulator, we have been able to stabilize that system using feedback. Actually, with this method we can not only stabilize the closed loop system but place the roots of the closed loop characteristic polynomial at arbitrary locations (*pole assignment*). This means that rather than placing the three roots at -1, -2 and -3, we could have chosen any three real numbers or one real number and two complex conjugate ones. We would then have been able to find real values for the three controller parameters that complete the (pole assignment) task.

6.5   Use of Feedback to Improve System Performance

In the previous section we demonstrated how feedback can be used to stabilize a given unstable plant. We used pole assignment to accomplish the task. In

Chapter 5 we saw that for second order systems (systems with transfer functions whose denominator is a polynomial of degree two), the location of the poles determines system tracking performance as well. Specifically, if we are interested in step responses, the location of the closed loop poles determines if the response is underdamped, overdamped, how large is the overshoot, how fast is the settling time, and so on. In the previous section we saw that by feedback we can arbitrarily choose the location of these poles. Consequently, we expect that feedback can also be used to improve closed loop system tracking performance.

Let us show how this can be accomplished in a specific example. Suppose that we are given a system with transfer function:

$$P(s) = \frac{1}{s(s+1)}$$

and corresponding differential equation:

$$\frac{d^2y(t)}{dt^2} + \frac{dy(t)}{dt} = f(t)$$

This is the transfer function of a mass-dashpot system (equation (2.8), $k = 0$). Since one of the poles of the system is $s = 0$ and lies on the $j\omega$ axis the system is unstable. Our first priority then is to stabilize the system (using feedback). In addition, we would like to address tracking performance issues related to step inputs. We would like the step response to be "fast," but to also have "small overshoot." From our discussion in Chapter 5 we already know that these are "conflicting" objectives. Let us use the feedback configuration of Figure 6.14.

Figure 6.14
Feedback Configuration with Gain Controller

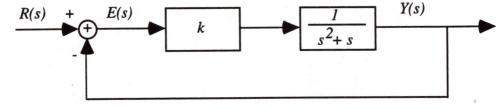

The closed loop transfer function can be obtained from expression (6.8) and is given by:

$$T(s) = \frac{k}{s^2 + s + k} \qquad (6.16)$$

From Section 5.1 we know that if k is chosen to be positive, then the closed loop system will be stable. It is also true that this transfer function is precisely of the form (see Section 5.2):

$$\frac{d^2y(t)}{dt^2} + 2\zeta\omega_n\frac{dy(t)}{dt} + \omega_n^2 y(t) = \omega_n^2 r(t)$$

($r(t)$ is the unit step) where for $k = 0$, we can immediately see that:

$$k = \omega_n^2 \quad \text{and} \quad 1 = 2\zeta\omega_n \tag{6.17}$$

It follows from (6.17) that the damping ratio ζ can be expressed as:

$$\zeta = \frac{1}{2\omega_n} \quad \text{or} \quad \zeta = \frac{1}{2\sqrt{k}}$$

By choosing k, we can affect both the undamped natural frequency ω_n and the damping ratio ζ. Recall that by making ω_n large, we make the system respond faster. This, however, would imply that the damping ratio will be made smaller (i.e., large overshoot). On the other hand, if we keep k (or ω_n) small, then the damping ratio ζ becomes larger and the overshoot would be reduced. These are contradictory requirements, which mean that we cannot make the system both fast *and* with small overshoot. We must compromise. So by choosing a large enough k we can have a reasonably fast response and not too large an overshoot.

Suppose that we choose $k = 1/4$. This implies that $\zeta = 1$. The step response (zero initial conditions) is shown in Figure 6.15 (solid curve). If $k = 1/2$, then $\zeta = .707$, (the step response is again displayed in Figure 6.15 as a dashed line). Finally, if $k = 1$, then $\zeta = .5$ (the step response is shown in Figure 6.15 as a dashed-dot line).

Figure 6.15
Step Responses
for System
(6.16)

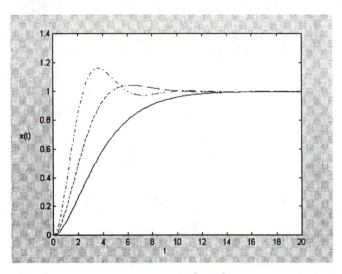

Comparing the three responses, we see that the step response performance exhibits the characteristics expected. The system is stable. Furthermore, as *k* is increased the response becomes faster, but at the same time the overshoot increases. In choosing the appropriate value we may select the second response (*k* = .5) as the compromise. It is important to remember that these changes in closed loop system performance are achieved through the use of feedback.

This is one specific example in which we see how the feedback structure allows us to improve the "closed loop" performance of dynamic systems using a simple controller. Just as we saw in system stabilization, a more complex controller will provide more degrees of freedom for design. These are topics that can be found in more advanced books on control (see Douglas, 1972, Ogunnaike and Ray, 1994 and Ogata, 1997).

6.6 Simulink Block Diagram with Feedback

In the previous sections we discussed how feedback can be used for system stabilization and performance improvement. We also suggested a way of computing the step response of the closed loop system. Namely, we first compute explicitly the closed loop transfer function and then solve the corresponding differential equation either analytically or using SIMULINK. If we choose to use SIMULINK, we construct the simulink block diagram as discussed in Chapter 4. SIMULINK is a very versatile software package that allows us to compute the closed loop step response directly. In particular, we do not have to first explicitly compute the closed loop transfer function. What we do instead is construct a simulink block diagram showing the feedback interconnections and have SIMULINK compute the closed loop transfer function and step response. This capability may not "buy" much for the simple feedback loop we have been working with, but it is invaluable for more complicated feedback interconnections (see Chapter 7). We now show how to construct such a simulink block diagram for the example in Section 6.5.

Consider the feedback system depicted in Figure 6.14, and assume we would like to compute the closed loop step response when the controller is $k = 1/4$. We first invoke SIMULINK and open up a new "untitled" window as explained in Chapter 4. The subsystem blocks we will need are the following: (1) from the "**Sources**" sub-library a "Step," (2) from the "**Sinks**" sub-library a "Scope," (3) from the " **Continuous**" sub-library *two* "Transfer Fcn" blocks and (4) from the "**Math**" sub-library a "Sum" block. These blocks are connected in exactly the same way as explained in Chapter 4 to form the feedback diagram depicted in Figure 6.14. The only new item is the connection of the "Sum" block. The "Sum" block can have several input ports and one output port and is used to form the algebraic sum of several signals. Its default setting is the sum of two signals and since in our example we need to form the difference, we need to modify it. We do this by opening its dialog box and changing the second "+" to a "-" as shown in Figure 6.16.

Figure 6.16
Dialog Block for "Sum"

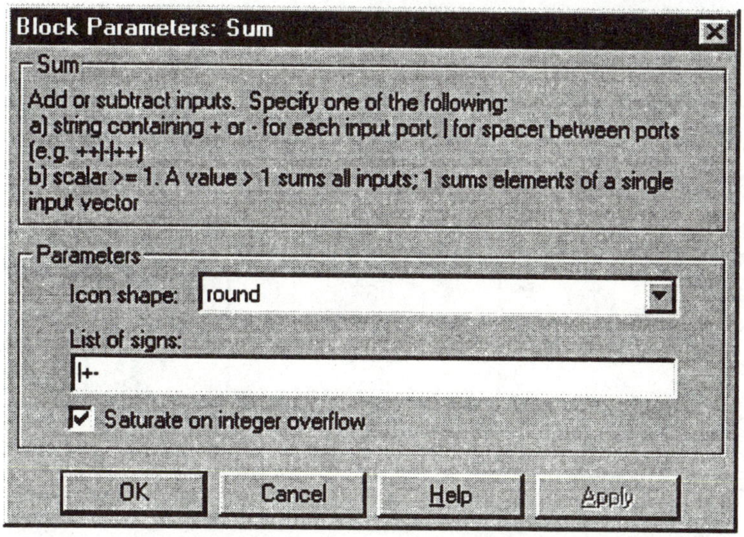

The "Sum" block is connected in the following manner: The output port of the "Step" block is connected to the side input port of the "Sum" block. Its output is connected to the "Transfer Fcn1" block of the controller. The output of the controller "Transfer Fcn1" block is connected to the input port of the system "Transfer Fcn" block. The output of the system "Transfer Fcn" block is connected to the input of the "Scope" block. By placing the pointer (cursor) on this last interconnection arrow and while pressing "Ctrl" key on the keyboard clicking and dragging the mouse, a new interconnection line appears with an arrowhead tip.

Figure 6.17
*Simulink Block
Diagram for
the Feedback
Interconnection
of Figure 6.14*

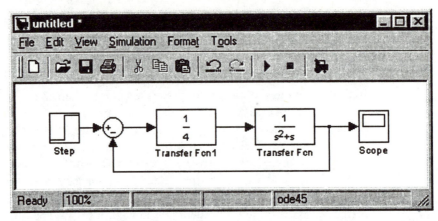

If we place the pointer at this tip the cursor turns into cross hairs. If we now click and drag the mouse the interconnection line can be extended to the bottom input port of the "Sum" block. After completing this operation and entering the appropriate data in the controller and system "Transfer Fcn" blocks, the simulink block diagram will look like that in Figure 6.17. We can now set the simulation parameters and run the simulation as explained in Chapter 4.

6.7 Exercises

Problem 1   Compute the overall system transfer function for the series interconnection shown below:

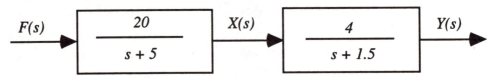

Is the overall system stable?

Problem 2   Compute the overall system transfer function for the series interconnection shown below:

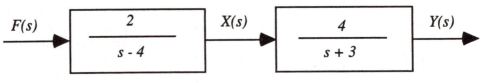

Is the overall system stable?

Problem 3   Compute the overall system transfer function for the feedback interconnection shown below:

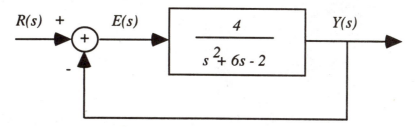

Is the closed loop system stable?

Problem 4   Compute the overall system transfer function for the feedback interconnection shown below:

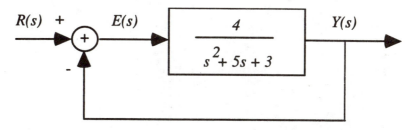

Verify that the closed loop system is stable. Use SIMULINK to compute the step response of the overall system (i.e., $r(t)$ is a unit step) when the initial conditions are zero ($0 \le t \le 20$).

Problem 5   Compute the overall system transfer function for the feedback interconnection shown below. Is the closed loop system stable?

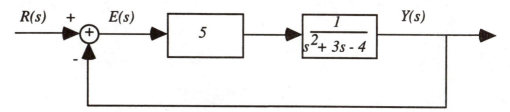

Problem 6   Consider the feedback interconnection shown below:

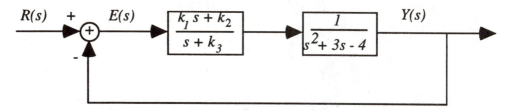

Find values for the controller parameters k_1, k_2 and k_3 that place the three closed loop poles of the system at locations -2, -4 and -5. Use SIMULINK to compute the step response of the overall system (i.e., $r(t)$ is a unit step) when the initial conditions are zero ($0 \leq t \leq 20$).

7 A COMPUTER-CONTROLLED MODEL CAR

7.1 Automatic Control of a Physical System

In the first six chapters we talked about mathematical models of dynamic systems, differential equations, methods of solution and digital computer simulations and introduced the concepts of stability, performance and feedback. We applied our results to dynamic models of systems and made a number of observations. However, our discussions were held at an "abstract" level because we have not as yet seen implementations on actual systems. In this chapter we will take one more step and demonstrate how the theory developed works on real systems. In fact, this methodology has been and continues to be successfully applied for the automatic control of a great variety of physical systems, from aircraft and spacecraft to robotic manipulators, chemical plants and automotive systems. We emphasize this point by focusing our attention in this chapter on one specific physical example and demonstrating how the theory developed can be applied for the automatic control of a model car.

The choice of an experimental "test-bed" for our automatic control implementation took into account a number of issues. We were interested not only in the experimental facility but in the specific experiments as well. Our considerations included the following:

- our primary concern was to choose an experimental facility that could be used for interesting and exciting experiments

- we wanted experiments that were sophisticated enough to demonstrate the power of our approach and yet simple enough so that they could be understood and appreciated

- we also wanted an experimental facility that would be representative of how automatic control is currently implemented in industry and would be relatively inexpensive to replicate

We chose what, in our view, is a system that meets all of these requirements. The experimental test-bed is a computer-controlled model car, and the experiment is one in collision avoidance. Our philosophy in this book has been to present a view of automatic control where theory, computer simulations and experiments are used together to solve engineering problems. It reflects how automatic control is practiced and it should be the manner in which automatic control is taught. This is the reason for including discussions about implementation in the book. Students here at the University of Massachusetts will have the opportunity to observe and participate in this experiment during class. Our hope is that students at other locations will also have a similar experience, perhaps with a different system. In any event, the written material presented in this chapter can be used as a "case study."

On several occasions throughout the book we have motivated our work by employing control problems associated with driving a car. This collision avoidance experiment is somewhat futuristic, since at this time cars, roads and highways are not equipped to support "automated driving." However, there is currently a great deal of research interest in "intelligent vehicle highway systems" (*IVHS*) and it is inevitable that in the near future we will begin to see commercially available systems for automated driving. Cruise-control is one of the few widespread commercially available automatically controlled systems to assist the operator in driving a vehicle. The "next generation" of such a system that we are likely to see implemented is referred to as intelligent cruise-control (*ICC*), which is specifically intended for highway driving. With additional sensors that provide information about the relative location and speed of neighboring vehicles, such a system would automatically maintain a safe distance from the vehicle directly ahead. In addition to throttle control, the system would incorporate the automatic application of the "brake." The computer-controlled model car discussed here is a "laboratory version" of such a system. The model car, built at the University of Massachusetts in Amherst, is called CIMCAR, which stands for Computer Intelligent Model CAR. The prototype CIMCAR-1 is shown in Figure 7.1.

Figure 7.1
The Computer-Controlled Model Car CIMCAR-1

The main components of CIMCAR-1 are:

(1)   the chassis, wheels and axles

(2)   a DC-motor

(3)   four rechargeable batteries

(4)   a sonar position sensor

(5)   an 8051 microcontroller

(6)   a digital-to-analog converter

(7)   a power amplifier

Batteries supply power (via the power amplifier) to the DC-motor, which in turn spins the rear wheels through a gear system. If the polarity is switched, the car moves backward. The batteries also supply power to the electronic components. Since it has its own power and a microcontroller on board, the car can move freely without being tethered to anything. At this stage CIMCAR-1 has very limited functionality. Since the front wheels are locked into position, it can only move in a straight line (future versions will have automatic steering capability as well). The sonar sensor provides relative position measurements between the car and any object ahead. Currently, in view of these limited capabilities, CIMCAR-1 can only be used for two types of experiments:

- the car can move toward an object and stop at a specified distance away from it

- the car can follow an object moving in a straight line at a specified distance behind it

These experiments are more than adequate to demonstrate all the theoretical concepts presented earlier in an innovative and exciting way.

7.2   A Transfer Function Model for CIMCAR-1

In our approach to automatic control, our first priority is to obtain a dynamic characterization of the system to be controlled. In Chapter 2 we talked about how physical laws can be used to develop models for physical systems. We also pointed out that such dynamic models can be obtained by performing some specially designed experiments. We did not elaborate on this matter earlier, but deferred discussion until now. Consider CIMCAR-1. It is an electro-mechanical system and as such its operation is governed by mechanical laws (Newton's laws) as well as electrical laws (Kirchoff's laws and others). One can construct dynamic models by referring back to "first principles" and laws from physics. Here we present a different approach. We will develop the system transfer function directly, using a three-step process that involves "experiments." Some of the technical details presented in this section may be too advanced for certain readers. In such a case, the reader should take the transfer function developed as given and proceed.

STEP 1 From an input-output viewpoint, we can think of our CIMCAR-1 as having one input (the voltage level supplied to the DC-motor that makes the car move) and one output (the position of the car from an object in front of it, say the wall). A conceptual block diagram of this system is shown in Figure 7.2.

Figure 7.2
Conceptual Block Diagram of CIMCAR-1

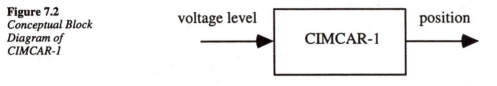

Our objective is to develop the dynamic relationship between the voltage level supplied to the DC-motor and the position of the car. Suppose for a moment that we have a *linear* relationship between position and voltage level. This seems like a reasonable assumption, since if we double the voltage level we would expect the speed of the car to double and so will the distance traveled. Clearly, our system is *time invariant*. This means that if we apply a certain voltage for a given time interval, the car will move a certain distance and if we repeat exactly the same experiment the next day, we see exactly the same results. In view of this linearity and time invariance, it is reasonable to propose a system model given in terms of a linear differential equation with constant coefficients. Equivalently, it could be given by some transfer function $P(s)$. In other words, if $v(t)$ is the voltage applied to the DC-motor (in volts), and $x(t)$ is the resulting position of the car (in inches), then we would have:

$$X(s) = P(s)V(s)$$

where $P(s)$ is some unknown transfer function. As a first step in this "plant identification" scheme, we postulate a transfer function model for our system. Assume a $P(s)$ and a corresponding differential equation with unknown parameters a, b, and K:

$$P(s) = \frac{K}{s^2 + as + b} \qquad \frac{d^2x(t)}{dt^2} + a\frac{dx(t)}{dt} + bx(t) = Kv(t)$$

This makes sense in view of our work in Chapter 2 on mechanical systems. The actual structure of this transfer function (and corresponding differential equation) will be dictated by the experimental results of Step 2 of this procedure. Our task then becomes one of trying to identify the appropriate values of these unknown parameters.

Here we mention a very important property of systems that are modeled by *stable* transfer functions $P(s)$ ($P(s)$ could have a factor s in the denominator). One can show that if the input to such a system is a sinusoidal signal of some frequency, the output in "steady state" will also be a sinusoid with the same frequency (see Ogata, 1997). Specifically, suppose that the input is $v(t) = V \sin(\omega t)$, where V is the amplitude and ω the "radian frequency" of the sinusoid. The output in steady state will have the form $x(t) = A \sin(\omega t + B)$. In fact, one can show that $A = |P(j\omega)|V$, where $|P(j\omega)|$ is the magnitude of the complex quantity $P(j\omega)$ ($j = \sqrt{-1}$). This means that $|P(j\omega)| = A/V$, or that the magnitude of $P(j\omega)$ is the ratio of the steady state output amplitude to the input amplitude.

STEP 2   In this step we suggest a set of experiments that can be performed on the system in order to "identify" the structure and specific parameters in the postulated transfer function model. The idea is the following:

(a)   apply a sinusoidal input voltage signal $v(t)$ to drive the DC-motor

(b)   measure the amplitude of the steady state sinusoidal output signal $x(t)$ and store the results

(c)   repeat steps a) and b) using a number of different voltage signals

In Step 1 we pointed out an important property of systems described by stable transfer functions dealing with sinusoidal inputs. For an input $v(t) = V \sin(\omega t)$, the units of V are volts, while the units of ω are radians per second. The larger the value of ω the faster the oscillations become, the larger the amplitude, the wider the voltage swings. We have postulated that CIMCAR has a transfer function model and as such it should exhibit this property. Indeed, we can observe such steady state performance when we excite the car's DC-motor with a sinusoidal voltage signal. The car swings back and forth in a sinusoidal fashion and the steady state output looks like $x(t) = A \sin(\omega t + B)$. We assume the output is a sinusoid and repeat this experiment for a number of different values ω (say ω_i for $1 = i = N$, with V kept fixed). The corresponding amplitudes of the output "sinusoids" (A_i for $1 = i = N$) are measured and recorded. From these data we compute the ratios A_i/V.

STEP 3   In this step we use the data recorded from the experiments in Step 2 to generate specific values for the parameters in the postulated model (i.e. find appropriate values for a, b and K).

The method presented here for this procedure involves trial and error and the idea is the following: pick a specific set of values for a, b and K that result in a particular stable transfer function $P(s)$ ($P(s)$ could have a factor s in the denominator). If we excite this system with a sinusoidal input, the steady state output will also be a sinusoid with the same frequency. Furthermore, the magnitude of $P(j\omega)$ is the ratio of the steady state output amplitude to the input amplitude. For each frequency ω_i from Step 2, we compare the magnitudes of $P(j\omega_i)$) with the experimental measurements A_i/V. If the two "match" (i.e., they are approximately equal) for all input signals, then we say that the system transfer function for the car is equal to $P(s)$. If not, we change the values for a, b and K, or even the structure of $P(s)$ and try again. This may seem rather cumbersome, but techniques do exist that facilitate this process. In the end a set of values for a, b and K is identified for which the computed output responses "match" the experimental data for the input signals used. Since the two match, we would have found a transfer function that for this set of inputs behaves in a similar manner as our model car. We can then say that this transfer function is a model for our system.

The result of this three-step identification procedure is the development of a transfer function (or equivalently a differential equation) for our system from experimental data. This method for constructing a system model is completely different from the one encountered earlier, because at no time during this process did we have to write down laws from physics in order to develop the model. Of course, our system will obey these laws as it operates. Clearly the model is an approximate one, but approximations would also have been needed, had we employed a method based directly on physical laws.

We now explain how this three-step plant identification procedure was applied to the CIMCAR-1. The input signal was a sinusoid of amplitude 5 (volts) and frequency ω_i, $1 = i = 17$. As we mentioned, in steady state the amplitude of the output signal is a sinusoid with corresponding amplitude A_i (inches). The frequencies ω_i, the measured amplitudes of the output signals A_i and the ratios A_i/V are given in Table 7.1, where $V = 5$. Now, we know that there is a "dead-zone" present in the DC-motor that was measured to be approximately ± 1 volt. This nonlinear phenomenon presents itself in this and many other applications. What we see for CIMCAR is that if we apply less than 1 volt to the DC-motor, the wheels not move. If the voltage is increased beyond this value the wheels begin to move. Clearly this is a nonlinear effect, since if the system had been linear and we had applied, say, 1.4 volts to the DC-motor, it would move at twice the rate as for .7 volts, and this does not happen. Therefore, it is appropriate that we take into account this nonlinear effect. We will show that one way in which this can be done is by increasing the "gain" in our transfer function model. This seems reasonable because the presence of a dead zone impedes the motion of the car. We should always remember that our system is actually nonlinear and that the transfer function, developed, will only be a linear model that is an approximate characterization.

Rather than plotting the values A_i/V versus ω_i for all $1 \leq i \leq 17$, it is customary to plot $20 \log_{10}(A_i/V)$ versus ω_i, where the x-axis is in "log-scale" (these are called *Bode Plots*, see Ogata, 1997). The values $20 \log_{10}(A_i/V)$ are in "decibels" (abbreviated "db"). Figure 7.3 shows the results. From these data, an appropriate structure for the transfer function model is:

$$P(s) = \frac{K}{s(s + a)}$$

The explanation of this fact is beyond the scope of this book (see Ogata, 1997, for more details). By trial and error we identify values for K and a that "fit" the experimental data. We chose $K = 20.8$ and $a = 3.8$.

Frequencies (rad/sec)	Output amplitudes (in)	Ratios
ω_i	A_i	A_i/V
.4398	60	12
.6283	42.75	8.55
.7540	34.75	6.95
1.0053	26.75	5.35
1.2566	20.85	4.17
1.5080	17.15	3.43
1.6336	15.6	3.12
2.0106	12	2.4
2.2619	10.5	2.1
2.5133	9.65	1.93
2.7646	8.25	1.65
3.0159	7.5	1.5
3.2673	7	1.4
3.5186	6	1.2
3.7699	5.5	1.1
4.0212	5	1
4.1470	4.5	.9

Table 7.1 Experimental Data from CIMCAR-1

Figure 7.3
*Bode Plot of
Experimental
Data for
CIMCAR-1*

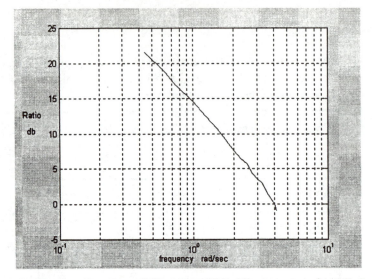

Figure 7.4 contains the plot of 20 $\log_{10}(|P(j\omega)|)$ versus ω for all frequencies (dotted line), superimposed on the corresponding plot of experimental data (solid line). Recall that $|P(j\omega)|$ indicates the magnitude of the complex quantity $P(j\omega)$. One can see that there is agreement between the transfer function model and the experimental data.

Figure 7.4
*Experimental
Data and System
model P(s).*

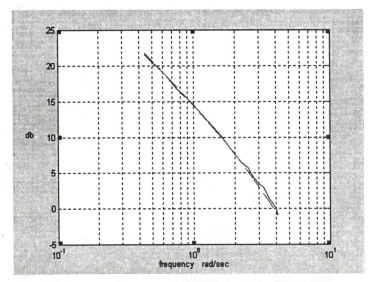

Let us now address the question of the dead zone. We use SIMULINK and its ability to simulate the dead-zone phenomenon to develop a model that takes it into account. We build a SIMULINK block diagram with a sinusoidal source block (found in the "Sources" subsystem library), a dead-zone block of size ±1 (found in the "Nonlinear" subsystem library), a transfer function block and a graph block. We choose the amplitude of the input sinusoid to be 5 and run several simulations with the frequencies used in the experiments (see Table

7.1). The transfer function block has transfer function $P(s) = K20.8/(s(s + 3.8))$ where K is adjusted so as to generate an output amplitude that matches the one experimentally measured. At the end of this procedure the linear transfer function model CIMCAR-1 is modified to be:

$$P(s) = \frac{28.5}{s(s + 3.8)}$$

The corresponding linear differential equation model is:

$$\frac{d^2x(t)}{dt^2} + 3.8\frac{dx(t)}{dt} = 28.5v(t)$$

where $v(t)$ is the voltage input to the DC-motor and $x(t)$ is the position of the car. Therefore, a more accurate dynamic description for CIMCAR is that shown in Figure 7.5. The model is made up of two components in series, a nonlinear component, which is the dead zone, and a linear transfer function.

Figure 7.5
A Nonlinear Model for CIMCAR-1

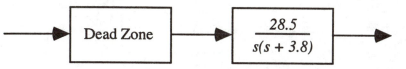

In the next section we will assume that the dead zone is actually not there and proceed to design controllers using just the linear transfer function model. Clearly this introduces "approximations" in our analysis and design, but there is ample justification for doing so: First, working with a linear model greatly facilitates the analysis and the design and allows us to gain a great deal of insight into the problem. Second, if the dead-zone effects are rather small (as in this case), accuracy will not be compromised too much. Third, the controllers we finally implement on the system will include a "dead-zone fix," which is intended to counter the effects of the dead zone.

7.3   A Collision Avoidance Experiment

In the previous section we described how to develop a transfer function model for CIMCAR-1. In this section we specify a collision avoidance task for CIMCAR-1 and formulate a solution that involves feedback. The collision avoidance task is defined as follows:

COLLISION AVOIDANCE TASK   The car should start eight feet away from a wall, move toward the wall and come to a stop three feet away from it. The performance specifications are:

- the car should be at 3 feet plus-or-minus 3 inches within 3 seconds

- the car should come to a stop 3 feet away from the wall without "overshooting" this target and certainly without crashing into the wall

Since the task is to be performed "automatically," the car must know at all times its distance away from the wall. This requires the use of a position sensor that would measure this distance and make it available to the car (*feedback*). The position sensor chosen for the car is a sonar sensor like the one found in auto-focus cameras. The sensor operates in the following manner: It periodically sends a "ping" (sound wave) and measures the time it takes for its reflection to reach it. Since we know how fast sound travels through air, the car can compute its relative distance from the wall.

If we assume that the position measurements are made available to the car, the next problem that must be addressed is to figure out how this information should be used, in order for the car to automatically complete the given task. The car moves by applying a voltage to the DC-motor, so the car must "figure out" what this voltage should be. One feedback scheme that looks promising is the one depicted in the conceptual block diagram in Figure 7.6.

Figure 7.6
A Conceptual Feedback Solution for the Collision Avoidance Task

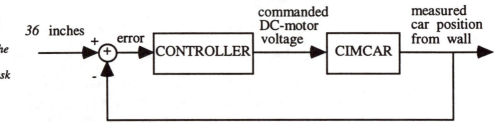

The solution we suggest does the following: The car starts at 96 inches away from the wall. It should be 36 inches away (reference position). The difference between the measured position and this reference creates an error. This error is used by the controller to develop a voltage command for the DC-motor, so that this error is driven to zero.

If we then decide to employ this feedback structure for our solution, the next priority is to construct an appropriate controller. The simplest such controller is just a "gain" controller. Specifically, the error signal is multiplied by some gain k and the product is the voltage command for the DC-motor. The scheme makes sense: the bigger the error, the larger the voltage applied to the DC-motor. A transfer function block diagram for this solution using the CIMCAR transfer function model is given in Figure 7.7.

Figure 7.7
Transfer Function Block Diagram for CIMCAR-1

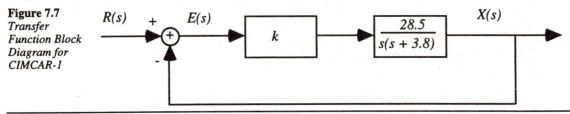

We can use this transfer function block diagram to analyze the dynamic behavior of our feedback system and, more specifically, to choose an appropriate value for k (i.e., design the controller).

Our first priority is to pick a value for k that makes the closed loop system stable (i.e., all signals should remain bounded). From our work in Chapter 6, we know that this is determined by the stability of the closed loop characteristic polynomial. Following the notation in Section 6.3 we have:

$$P(s) = \frac{n_p(s)}{d_p(s)} = \frac{28.5}{s(s + 3.8)}$$

$$C(s) = \frac{n_c(s)}{d_c(s)} = \frac{k}{1}$$

and

$$\phi(s) = d_p(s)d_c(s) + n_p(s)n_c(s) = s(s + 3.8) + 28.5k = s^2 + 3.8s + 28.5k$$

From this we see that the closed loop system will be stable if, and only if, the controller gain k is a positive number. This agrees with our physical intuition. Negative voltage to the motor makes the car move toward the wall while positive voltage makes the car move away from the wall. If the car is at 96 inches from the wall and must go to 36 inches it must move towards the wall. This implies the voltage to the motor must be negative. If k were negative, this voltage would have been positive. The car would then move away from the wall.

Our feedback solution requires our car to move from 96 inches away from the wall at time $t = 0$, to 36 inches away. This implies that the reference input is a step of size 36 and the initial position is 96. Equivalently, we can think of this as having the car go from 0 inches to -60 inches (i.e., a step of size -60 starting from the zero initial position). The dynamic behavior of such systems was considered in Chapter 5. Furthermore, if we look closely at the feedback diagrams in Figures 7.6 and 7.7, we realize that the proposed solution requires that this move be done in "zero time." This is certainly unrealistic; a more reasonable reference trajectory would be specified by a function that starts at 96 (at time $t = 0$) and gradually decreases to the value 36 in some time interval. In keeping with the introductory nature of this book, we will not pursue this further. However, we will see that we can do quite well with the proposed solution.

So the reference trajectory is chosen to be the step function:

$$r(t) = \begin{cases} 0 & t > 0 \\ 36 & t \geq 0 \end{cases} \tag{7.1}$$

The overall transfer function for the system from equation (6.8) is:

$$T(s) = \frac{n_p(s)n_c(s)}{d_p(s)d_c(s) + n_p(s)n_c(s)} = \frac{28.5k}{s^2 + 3.8s + 28.5k} \tag{7.2}$$

The corresponding differential equation is:

$$\frac{d^2x(t)}{dt^2} + 3.8\frac{dx(t)}{dt} + 28.5kx(t) = 28.5kr(t) \tag{7.3}$$

which is of the form (see Section 5.2):

$$\frac{d^2x(t)}{dt^2} + 2\zeta\omega_n\frac{dx(t)}{dt} + \omega_n^2 x(t) = \omega_n^2 r(t)$$

Note that our initial conditions are: $x(0) = 96$, $x'(0) = 0$ (initial position of the car is 96 inches away from wall and initial velocity is zero). Our system is second order; in Chapter 5 we talked about stability and step responses of such systems. In particular, we have:

$$\omega_n^2 = 28.5k \qquad 2\zeta\omega_n = 3.8$$

which implies that:

$$\omega_n = \sqrt{28.5k} \qquad \zeta = \frac{1.9}{\sqrt{28.5k}} \tag{7.4}$$

From Chapter 5 we know that the values of ω_n and ζ determine the step response of the system. Specifically, if $\zeta = 1$, the system is overdamped and if $0 \leq \zeta < 1$ it is underdamped. From above, it is clear that the value of k directly specifies the values of ω_n and ζ. Recall also that an underdamped system has overshoot in the step response and that the larger the value of ω_n, the faster the response. Therefore, in view of our performance specifications, which require no overshoot, it is evident that our controller design should make the system "overdamped."

We are now ready to use the theory developed in Chapters 5 and 6 and design a controller that meets the performance objectives set for this collision avoidance task. For this system it is relatively straightforward to analytically determine a range of values of k that make the system response overdamped (i.e., $\zeta = 1$). From equation (7.4) the range is:

$$k = .1266 \approx .13$$

We will present the results from three values of k, where one of the designs will be for an underdamped system. Even though this will not be our final design, we include it here for illustrative purposes. Our objective here is to demonstrate agreement between theory, digital computer simulations and experiments.

DESIGN 1   Suppose that $k = .5$. Then directly from (7.4) we have that:

$$\zeta = .50 \quad \text{and} \quad \omega_n = 3.78$$

Since ζ is in the range 0 to 1, the step response will be underdamped and will show overshoot. Clearly, this does not meet our requirements, a fact verified by carrying out a simulation of the system. Since the initial conditions are not zero, we will use a "State-Space" subsystem block (as in Example 4.4) to represent our system in a simulink block diagram (see Figure 7.8).

Figure 7.8
Simulink Block Diagram for Equation (7.3)

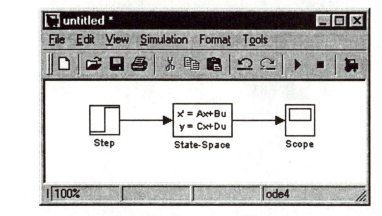

The final value in the "Step" block is set to 36. The "State-Space" dialog box is completed as shown in Figure 7.9.

Figure 7.9 *State-Space Dialog Box for Equation (7.2)*

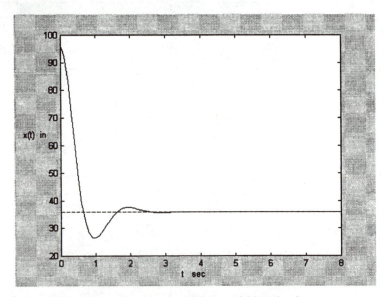

The step response is shown in Figure 7.10.

Figure 7.10
Step Response of Equation (7.3) for k = .5

DESIGN 2   Let us now choose the value $k = .1$. This would imply that:

$$\zeta = 1.13 \quad \text{and} \quad \omega_n = 1.69$$

Since ζ is bigger than 1, we know that the step response will be overdamped with no overshoot, but more sluggish than the first design because ω_n is now smaller. The step response of equation (7.3) with $k = .1$ is given in Figure 7.11. The parameter modifications for the State-Space dialog box in this case are: $A = [0 \ 1;-2.85 \ -3.8]$, $B = [0; 2.85]$, with the rest remaining as in Figure 7.9.

Figure 7.11
Step Response of Equation (7.3) for k = .1

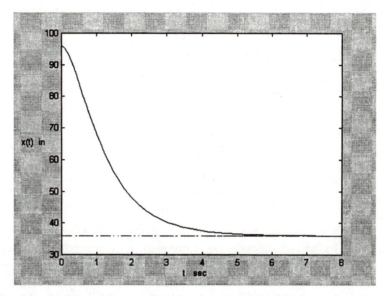

Even though we were able to solve the overshoot problem, we see that the car does not move fast enough.

DESIGN 3   Clearly, the value we need for k is somewhere in between the two previous values. Let us now set $k = .13$. This results in (the practically critically damped case):

$$\zeta = .99 \quad \text{and} \quad \omega_n = 1.92$$

The parameters for the State-Space dialog box in this case are: $A = [0 \ 1;-3.705 \ -3.8]$, $B = [0; 3.705]$, with the rest in Figure 7.9. The step response is shown in Figure 7.12.

Figure 7.12
*Step Response of
Equation (7.3) for
k = .13*

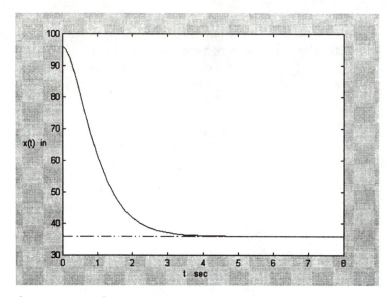

Both performance requirements have been met, since the response shows no overshoot and completes the move within three seconds.

These are the results predicted from our "linear" analysis and controller design. In the next section we show results from experiments with the CIMCAR, and we will see that they pretty much agree with those predicted by the theory. We cannot expect that there will be perfect agreement, because the dynamic model for the CIMCAR is only a linear approximation of the true model (recall the dead zone). Since SIMULINK does give us the ability to simulate nonlinear system models, we will use it to perform a "nonlinear" simulation and compare the results with the experimental data. It will be very encouraging to see the agreement between theory, simulation results and the experimental data.

7.4   Experimental Results

We have just completed a controller design, based on the theory developed earlier in the book. This design was checked by simulation. Our next priority is to implement these three different controllers on the car itself and *verify* that experimental results agree with those predicted by theory and demonstrated via simulation. The goal of control engineers is not only the paper or computer design of controllers but their successful physical implementation on "real" systems. It is also important to keep in mind that we must *validate* any theoretical controller design technique by demonstrating its successful implementation on physical systems. If we are not able to do this on a consistent basis, all our hard work involving physics and mathematics would be in vain!

In the previous section we have designed three simple controllers for the CIMCAR; we are now ready to physically implement them. Basically, there are two ways that this can be done. One can use either analog circuitry to implement a controller or a digital implementation. What this implies is that the controller will be implemented in software rather than hardware. Only those readers who have already done some circuit design can appreciate the difference. Electrical engineering students will learn more about this topic in the future. In the design of CIMCAR we opted to follow the digital route for a number of reasons. Perhaps one of the most important is the fact that this is how collision avoidance systems will be implemented on real cars.

Before we proceed, it is important to remember that the controllers for CIMCAR were designed based on an *approximate linear* transfer function model of the system. Recall that even though we mentioned that there is a dead zone present in the system, we neglected to take it into account when designing controllers. We alluded to the fact that the controllers we will finally implement on the car will include a dead-zone fix. We will discuss this issue now, as well as another concern that has to do with "protection" of the DC-motor.

There are a number of ways for us to address the dead-zone issue. Some take advantage of theoretical results that have been developed, but the approach we will suggest is practical. In order to reduce the effects of this dead zone, we have devised the following scheme to fix the problem: If the required DC-motor voltage v has a value $|v| = 1.5$ volts, do nothing to it. If v is in the range $|v| = .5$ volts, also do nothing (which in simulation implies that the voltage is set to zero). If however, $.5 = |v| = 1.5$, then provide the voltage 1.5 (or -1.5 as the case may be) to the motor. This is only one of a number of solutions that can be proposed and it is by no means the best. Our controller for CIMCAR is now comprised of the gain k, as well as this dead-zone fix. This controller is again implemented by a few lines of code on the 8051 microcontroller.

An additional concern is the protection of the DC-motor so that the voltage supplied to it is within the safe range. To accomplish this we implement a "protection" mechanism that limits the voltage to the range $-10 = v = 10$ volts. This means that if the controller commands that more than 10 (or -10) volts be supplied to the DC-motor, only 10 (or -10) volts are actually provided. This operation can again be very easily programmed with a few lines of code on the 8051 microcontroller.

At this stage of the design we have modified the initial controller by including the dead-zone fix and have included a protection mechanism for the motor. Before we implement the entire scheme on the car, it is important to go back and, with the help of SIMULINK, simulate the overall system. SIMULINK has the capability of including not only these but many other nonlinear effects in simulations. Specifically, in the "Nonlinear" subsystem library we can find blocks for a dead zone as well as a "saturation" (the protection mechanism). It also provides other blocks that allow us to construct our own block for the dead-zone fix. Using these blocks, we generate the following simulink block diagram (see Figure 7.13). In addition to the dead-zone fix and the saturation blocks, we have included a dead-zone block to represent the dead zone present in our system. It should be apparent that this is a more accurate description of our dynamic system.

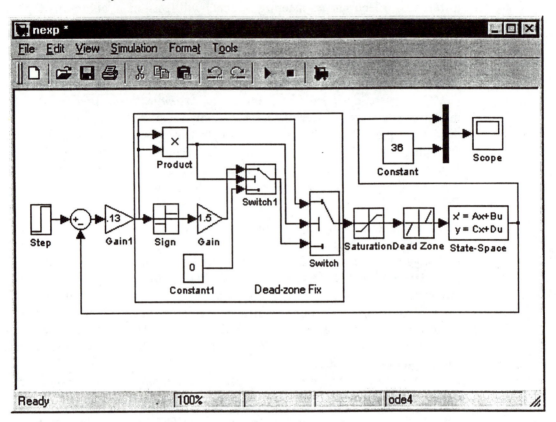

Figure 7.13 *Simulink Block Diagram with Dead-Zone Fix, Saturation and Dead Zone*

The plots in Figures 7.14, 7.15 and 7.16 give the simulation results (dashed lines) as well as the experimental results (solid lines). Figure 7.14 corresponds to a controller gain .5, Figure 7.15 to the controller gain .1 and Figure 7.16 to the controller gain .13. As one can see, there is great agreement between the experimental results and the nonlinear simulation results. These also agree "qualitatively" with the linear simulations (Figures 7.10, 7.11 and 7.12, respectively) and confirm the theoretical predictions. This agreement between theory, simulation and experiments is particularly important because it validates the strength and credibility of our approach.

Figure 7.14
*Experimental and
Simulation Results
for Controller k
=.5*

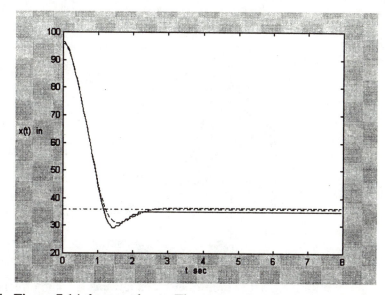

Notice in Figure 7.14 the overshoot. The car reaches the desired distance away from the wall but it "overshoots." The response is fast but it violates the performance specification of "no overshoot."

Figure 7.15
*Experimental and
Simulation Results
for Controller k
=.1*

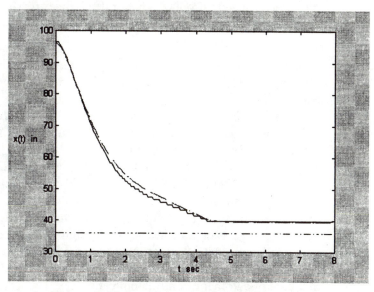

The response in Figure 7.15 has no overshoot but it is too slow. Clearly, a controller gain which is in-between these two values would give an acceptable response and this is shown in Figure 7.16.

Figure 7.16
*Experimental and
Simulation Results
for Controller k
=.13*

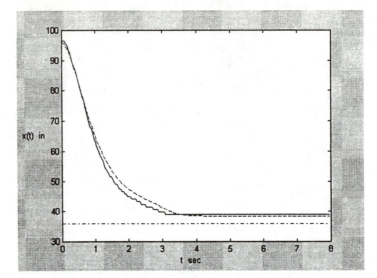

With the smaller controller gains (.1 and .13) the effects of the dead zone are more pronounced, as the car settles at 39 inches away from the wall. However, this controller design meets the requirements we had set for the task. Had our requirements been more stringent, we would have needed to modify the controller gain and introduce a better dead-zone fix. It is important to note that the linear design gave us a good controller and provided a lot of insight.

This example gives a very good picture of what automatic control is all about. We hope is that we have successfully communicated this to you. The proposed solution uses feedback in a decisive way to guarantee that the performance requirements for our collision avoidance task are met. An initial controller is designed based on a simple linear model of the CIMCAR, which meets the performance objectives in theory and simulation. This controller is then suitably modified by employing a dead-zone fix to address this nonlinear phenomenon. Computer simulations are again performed in SIMULINK, which now include the nonlinear effects. Experiments are then carried out to verify that the specifications are met on the actual system. In this example we were quite successful in meeting our performance goals. In some cases it may necessary to go back and iterate this design procedure (i.e., develop a better model, redesign controllers, etc.). The approach we have outlined in this book is very representative of how controllers are designed and implemented in a multitude of automatic control applications.

BIBLIOGRAPHY

1. Boyce, W. E., and R. C. DiPrima. *Elementary Differential Equations and Boundary Value Problems*, Third Edition. New York: J. Wiley, 1977.

2. Douglas, J. *Process Dynamics and Control, Volumes I and II*. Englewood Cliffs, NJ: Prentice-Hall, 1972.

3. Ogunnaike, B. A., and W. H. Ray. *Process Dymnamics, Modeling and Control*. New York: Oxford University Press, 1994.

4. Ogata, K. *Modern Control Engineering*, Third Edition. Englewood Cliffs, NJ: Prentice-Hall, 1997.

5. The MathWorks, *SIMULINK, Version 3*. Natick, MA: The MathWorks, Inc. 1999.

6. Bobrow, L. *Elementary Linear Circuit Analysis*, Second Edition. New York: Oxford University Press, 1987.

INDEX